Communication in Modern Social Ordering

Communication in Modern Social Ordering

History and Philosophy

Kai Eriksson

continuum

The Continuum International Publishing Group
80 Maiden Lane, New York, NY 10038
The Tower Building, 11 York Road, London SE1 7NX

www.continuumbooks.com

Library of Congress Cataloging-in-Publication Data
A catalog record for this book is available from the Library of Congress.

ISBN: HB: 978-1-4411-7126-9

Typeset by Pindar NZ, Auckland, New Zealand
Printed and bound in the United States of America

Contents

For Päivi and Justus

1

Introduction

The book at hand will investigate communication and its changing conditions with respect to the conceptualizations of the political community in the United States. It examines the communicative forms of constructing society from the nineteenth century onward. The central aim is to deliberate upon how the ideas relating to communication technologies ranging from the telegraph to computer networks have organized the conception of society and how they have played a part in the shaping, ordering, and governing of the society according to certain political aims and principles. While communication has structured the political community and its coherence, communication has also, in the same process, become organized as a concept itself. Together with the discourses in which communication has been taken as both the object and the means of governance practices, communication technologies have formed an environment wherein "communication" as such has become an object of systematic thought — that is, it has become a distinct theoretical concept and a social concern. It is in these discourses and practices that the inherent relationship between communication and control was forged, and this is what makes them the main focus for a "modern" history of communication — a history preoccupied with the idea of control. The book will analyze the relationship between historical forms and technologies of communication, on the one hand, and the constitutive, ontological dimension of communication, on the other, to the extent that this relationship is built up through — and has served as a conceptual means for understanding — the society and its central processes in the modern era. It thus sets out to examine how communication has generated coherence to the political community and how a certain coherence, on the other hand, has been a precondition for thinking about communication as a socially significant concept.

The development of communication technology has undoubtedly been one of the grand narratives of our time. In this narrative, the main concepts and images have almost invariably been based on different types of totality. From Samuel Morse's "one neighborhood" to Marshall McLuhan's "global village,"[1] the modes of thinking about the power of communication have been bound by the idea of a united whole.

1 One neighborhood: Samuel F. B. Morse, "Morse's letter to F. O. J. Smith," February 15, 1838, in *His Letters and Journals: Invention of the Telegraph, Vol. II*, ed. Edward Lind Morse (Boston: Houghton Mifflin, 1914), 85; Global village: Marshall McLuhan and Bruce Powers, *The Global Village: Transformations in the World Life and Media in the 21st Century* (New York: Oxford University Press, 1989).

Despite some notable exceptions, such as John Dewey and Charles Cooley for instance, it has been customary for "the moderns" to consider society and its communicative nature through a rather consistent conceptual model, often derived from the rapidly improving technological systems, including telecommunications, which facilitated rationalization, differentiation, and control. Together with related discourses and practices, these systems have formed a significant environment wherein communication as such has become a central concern, especially when it has been taken as both the object and the means of administrative practices.

The word telecommunication — the prefix *tele* dating from the Greek word τηλε, meaning far, remote, far away — came to mean long-distance communication that is based on electronic communication technology. However, although the word was not adopted internationally until 1932, telecommunications has a long history that predates its concept, originating in the first half of the nineteenth century. The book shows that the history of telecommunications and the related ideas and discourses can be divided into three stages, broadly corresponding to the eras of telegraphy, telephony, and computing respectively. Each of these eras can be considered to have sustained a different idea of communication and the communicative system, although many of the ideas also intertwined and were used in several different time periods. Thus, during the era of the telegraph, the conception of a living organism, inspired by the theories of Herbert Spencer and William Sumner, provided a popular model for the conceptualization of both communication and society. It was especially the idea of the nervous system that constituted the framework for the discourse of communication, understood as a means and basis for certain overarching political, commercial, and administrative projects. Since the turn of the century, as the telephone became a common device, the conception of communication progressively broke away from the organicist language and was tied to the metaphor of the machine and the idea of a rational, hierarchically controlled system. This took place in the shadow of the technocratic conception of communication primarily as a question of technique and social control, influenced by a theoretical tradition extending from Lester Ward through Frederick W. Taylor to the technocracy movement. Finally, in the postwar period when the new information and communication technologies assumed social prominence, the conception of communication was increasingly connected to systems controlled by the principles of feedback and circularity within a theoretical horizon formed by cybernetics and the systems sciences. It was Norbert Wiener and Claude E. Shannon, in particular, who laid the new communicative framework in which ideas and models of communication gradually left mechanical metaphors behind and entered the sphere of influence of information system analogies. The book deliberates upon how these conceptualizations have been intertwined and have become integral parts of the construction of a modern political community, both technologically and conceptually, and of the articulation of political order as a question of communication.

Communication is not, of course, totally assimilated into the techniques and systems that shape its forms of realization but constantly escapes their functionality: new modes of communication ensue from the coincidence of heterogeneous forces in the communicative event. It spawns new, unprecedented practices, develops ramifications in fields and directions calling for institutionalization, and endlessly yields interaction that is beyond the reach of governance. Communication systems constantly take new forms to meet emerging practices as communicative events constantly reconstitute the institutional conditions of these systems. Therefore, although it is always instrumental in relation to the ends it is connected with and whose functions it performs, communication is never purely external in the relationships it expresses. Rather, it can be seen as an effect of this network of relationships, the dynamics of which it expresses, but at the same time also the means through which the network in question functions and gets restructured, and which, thus, is not only the voice but also the mechanism of this process. This is what, from the perspective of the work at hand, is of importance: to think about how communication practices continued, broke off, reorganized, and reinterpreted not only the social production and sharing of conceptual identities but also the things shared, the shared tradition, and communality.

Thus, the analysis here is not centered exclusively on communication seen as technology — although this aspect cannot be cut off from the emergence of the modern communication and social order, but, on the contrary, constitutes its precondition. Rather, the emphasis is on the changes in and of communication during the period in question; that is, during what has commonly been understood as modernity. Modernity has often been referred to as the industrial order that broke away from the agricultural system based on relatively self-sufficient localities and strong local government. However, it also refers to the emergence of a particular way of conduct, defined by an obsession with rationalization, systematization, and control. Centralized order, consistent meaning and the closed whole represented the principal ideals and goals of this mode of conduct. The central purpose here is to discuss the relation between communication and specific communication technologies, which has been approached as a question of political community. Communication is not seen as dissociated from the rest of the societal processes but rather is mediated through different communication systems and schemes as well as through the various related discourses and concepts. The conceptions of communication that originate from these systems and arrangements become, for their part, intertwined with the political thought about the community, including questions related to the theoretical, administrative, and juridical concerns pertaining to that community.

The book thus deals with communication as both a tool and a concept. I will consider the way in which communication has engendered distinct formations or constellations of theories, technologies, and institutions through which a community has been able to contemplate itself, as it were, within relatively coherent frameworks. At the same time, however, communication has been taken as a political technology in the process of social ordering. I will consider how it has become a means of generating political cohesion and how, in the course of this, communication has also organized the discourses and imagery in and through which this aim could be given a consistent expression. Thus one can see that these two meanings are not separate but closely connected: communication as a technology was a major tool for engendering political unity in what seemed to be a fragmented social space, and the idea of a unified national whole served as the framework in which the diverse communication processes were able to be seen as a part of a highly systematic structure.

The word "communication" is derived from the Latin term *communicatio*, which, in turn, has its foundation in the word *communis*, meaning "common to several, or to all." The word is understood to have two basic meanings. According to the first meaning group, it signifies the action of imparting or transmitting something. In the fourteenth century, the word referred to the action of communicating something or of giving something to be shared. At that time, it referred mainly to material objects and discernible experiences such as heat, feeling, or motion. Around the middle of the sixteenth century, the use of the term extended to include the transmission or exchange of information, knowledge, or ideas, as for instance by means of speech or writing, and later by mechanical or electronic media. This specific sense of imparting or transmitting ideas or knowledge is probably to be found first in English in John Locke's *An Essay Concerning Human Understanding* in 1690. In the latter half of the nineteenth century through to the early twentieth century, the word was gradually connected to the science, practice, or system of transmitting information over long distances, especially by electrical means such as by telegraphy, telephony, or broadcasting. This is why this meaning of the word has, particularly in its modern sense, close connection to communication technology.

According to the other meaning group, the word communication relates principally to affinity, association, and participation. In the late fourteenth century, it was used to express the fact of having something in common with another person or thing, thus containing a sense of affinity or congruity. The use of the word was extended in the first half of the following century to interpersonal contact, social interaction, association, and intercourse. By the sixteenth century, the word was understood as the action of sharing something, mutual participation, or ownership. Thus it became possible, for instance, for J. Whitgift, in 1574, to speak about the banishment of a man "from the communication of prayer, of companie, and of all holie affaires," and later on for Thomas Jefferson, in his *Draft Declaration of Independence* in 1776, about "a communication of grandeur and of freedom" (*Oxford English Dictionary Online*). One of the Latin roots of the word is *commūnicāre*, meaning precisely to make common to many, to share, and to impart. Accordingly, "to communicate" thus meant, in around 1526, to partake in common, to share (*Chambers Dictionary of Etymology*).

Following this distinction, the communication theorist James Carey has identified two conceptions of communication: first, there is what he calls the transmission view of communication, being defined by terms such as imparting, sending, transmitting, or giving information to others. The core principle of this view, allegedly the most common one in Western culture, is that communication is basically about transmission of signals or messages over distance for the purpose of control. The idea of overcoming long distances became emphasized when the communication systems improved and technology became available for long-distance communication. Thus communication understood in the context of communication technologies has been typically viewed as being about forging connections, by means of separate messages, between detached individuals, locations, and people. It is organized as a linear line between two communicating parties. Second, there is what Carey has termed the ritual view of communication, in which communication is linked to terms such as sharing, participation, association, fellowship, and the possession of common faith. Here, the central idea is not organized around transmission and information, but around the intrinsic relationships and common roots of the concepts of commonness, communion, community, and communication: it is closely connected to the activity of creating something in common as a way of engendering and articulating our common world, of constructing shared representations through which this world could be made comprehensible.[2]

Most of the political significance that communication gained was connected to its potential in solidifying the community by providing both technical apparatuses and conceptual models toward this end. In this process, communication has been gradually recognized as the constitutive precondition for communal processes and social institutions. It is thus not only a tool of organization but also a language in which this organization can be brought up and on the basis of which it can be designed. Neither does it appear only as an object of representation. Rather, it appears as a locus for social orders to take shape, as a place in which those orders can be articulated, as well as an agent of their realization and as a factor that maintains and embodies their formation. In this way communication is originally connected with the notions of social co-being and community-building, even though the idea of communication as an action based on individual messages has promoted the view of it as an independent "product," detachable from its environment. However, communication is not only something isolated and separate. More to the point, it is about building the community through engendering communicative space. While the former, transmission view has arguably dominated the American thought of communication, it is particularly the latter one and its inherent notion that communication constitutes the community that provides the ontological angle employed in this book.

Communicating instances originally presuppose communication: they do not take place in relation to others as if already perfect subjects of communication — as was thought in the era of the telegraph as well as, although in a different way, at the time the telephone became common and the conception of communication was rationalized — but this relationship, communication, is exactly what enables the subject positions and the production of meaning. In the process of the rationalization of communication, this ontological dimension tends to become unimportant: communicating subjects' relationships to communication is seen on the basis of a model wherein a subject assumes an originality that it expresses by means of communication. Subjects here are conceived as not having an inherent,

2 James Carey, "A Cultural Approach to Communication," *Communication* 2, (1975): 1–22.

only an external, relationship between each other, and communication is the mediator between original intentions and the needs of interacting subjects. What has been lost in this conception is the fact that the subject of communication always already presupposes others as the precondition of communication: the other is not an external object of messages but more fundamentally constitutes the ontological condition of a communicating subject.

For an ontological point of view, communication is thus not just a vehicle for transmitting information between pre-existing subjects and for intensifying relationships that already exist. Rather, it enables and opens up something that would not be possible without it — a new sense of national cohesion or a feeling of togetherness, or a new idea of how diverse social processes might actually function together to form a highly sophisticated apparatus. It is surely not to be reduced to the transmission of singular messages as has been the case, indeed the rule, over most of the modern history of communication. Communication is also, and perhaps most fundamentally, the constitutive condition of the community and the social system, or, in general, what can be called as "the social."[3] And yet the linear, transmission view of communication is an integral constituent of modernity. This book tries to concentrate on what Carey seemed to ignore: although communication is not simply about transmission but rather about producing and maintaining the common world, it is the very idea of communication as transmission through which the modern world has been shaped and transformed. Thus, in a way, this book is about the ontology of transmission or mediation.

The constitutive dimension of communication can be seen as transcendental with respect to the day-to-day realization of communication, as it constitutes its precondition. Yet this transcendental dimension does not form any infinite, unchanging structure external to the very historicity of communication. The contours of the domain of communicability are drawn historically, marked and implemented by the means through which people have tried to capture the ever-evading secret of communication.[4] Because these means can be realized as historical means only, it is possible to locate some historical identities of the

3 The concept "the social" is understood here as the instance that allows us to recognize and define social activity *as* social. Similar definition applies to "the political" with regard to political activity. For a more contemporary meaning of the concept, "the social" can be conceived of as the articulation of a space of governmentality that is composed of heterogeneous and interconnected processes and programs linking the private with the public sphere. If the social is viewed in terms of the discourse on governmentality, then it is possible to identify the forms it has assumed both as the object and as the means of various policy measures and strategies. See Mitchell Dean, *Governmentality: Power and Rule in Modern Society* (London: Sage, 1999); Nikolas Rose, *Powers of Freedom: Reframing Political Thought* (Cambridge: Cambridge University Press, 1999); see Michel Foucault, "Governmentality," in *The Foucault Effect: Studies in Governmentality*, ed. Graham Burchell, Colin Gordon, and Peter Miller (Hemel Hempstead: Harvester, 1991). In line with this interpretation, my focus is not on institutions or political structures in themselves, but rather on the interweaving of different theories, programs, practices, and technologies in a given historical period, which together form an intelligible whole.

4 These projects, of course, are illegitimate: if communication can be thought of as having a character of its own, it resides precisely in its nature as denying or repelling any single essence. In any case, these conceptualizations express the necessity to take communication in itself as an object of thought, although these projects understandably cannot ever be successful in their goal of presenting what is ultimately unpresentable.

courses they have taken. A natural way to trace these is to study the mechanisms and areas of exclusion through which communication has been distinguished from what it is not — its outside — which has to be separated from communication in order for it to maintain a needed coherence. As elements of this outside, noise and multiplicity can be viewed here as the enemy of communication, something that can swamp the established order within a given system. These processes of exclusion from communication provide a possibility of tracing not only the historical configurations, patterns, and structures of communication but also, above all, the question of the *limits* of communication as it is organized historically.

Thus the book places the ontological framing of communication into a historical context with a view of making manifest the historicity of ontology. Ontology is not only about the ideas that some philosophers have formulated: it is also, and perhaps more important, about the historically changing frameworks that allow contemporaries to verbalize and express novel questions and ideas and in this way to open up a new relationship to the world in terms of communication. In this sense the book tries precisely to historicize the ontology of communication — the changing conceptual conditions for articulating the truth concerning what communication is. Thus it focuses on the way in which what was previously unthinkable becomes as intelligible and articulate. Furthermore, it investigates how the whole idea of what "communication" itself is about actually changes, thus making it possible to link up processes and entities in completely new ways. Therefore, the main concern is not how some notable thinkers have conceptualized communication. Rather, it is how the whole conceptual system in which communication came to be thought about changed over time and allowed radically new ideas and formulations of communication to emerge and become established.

By communication, I understand the concern, both theoretical and political, that achieved distinctiveness and prominence during the period studied here. While communication took form as a concept of its own, it also became institutionalized into scientific research and administrative practice. It should thus be understood broadly as referring to a historically changing set of interrelationships among a concept, a social phenomenon, different technological systems, diverse political programs, and ultimately certain images of society. If the matter is approached from the standpoint of politics, then the key dimensions of communication are defined by the mutual interaction and articulation of these factors. What is to be scrutinized, then, is the way in which communication is organized in relation to the political community, investigating the particular relationship each conception of communication creates for and within the political community: how does communication constitute the political community in the era of telecommunications?

It was only from the era of intraurban telegraph systems onward that communication assumed a truly positive character, developing into a productive force that represented the core of the modern society. It was this new positivity that made it an object of systematic thought: concurrently with the development of communication technology, a theoretical interest in its hidden mechanisms and its underlying laws arose, attempting to encapsulate them in a rational model in order not only to ensure the dependability of certain communication systems but also, and most important, to understand communication in its essence.[5] Thus, as telecommunications became established and its cultural implications visible, communication was conceptualized increasingly as an object of action, not just as a means to that end.

The demarcation between science and nature, subject and object, drawn from the idealized model of the natural sciences, led social scientists, in the second half of the nineteenth century, to analyze their objects as isolated and relatively coherent. The attempt to justify one's disciplinary approach in a way that would meet the criteria of scientificity, by defining its object as being real in some objective sense as well as being separate from the observer, had been established in the social sciences already by Émile Durkheim, and took firm root in its subsequent development in thought about "communication." This way of conceptualization has served some apparent aims. It has become much easier, for example, to speak about the "impacts" of certain forms of communication on different sectors of society. Similarly, processes such as design, implementation, or management of given types of communication technology can be organized as a systematizable part of a system of knowledge. In its modern history, communication and the communicative space in which it appears have been conceived of as a distinct conceptual whole, the characteristics of which have become more mathematical the more telecommunication systems have developed.

In this book, communication is viewed as neither homogeneous nor uniform, as it does not determine a subject that can be understood as unitary or composed of similar units. This is because it always takes place in some locatable situation that is made up of a given assemblage of social forces, and is thus never separate from the power effects actualized by way of this set of relationships. A central part of the constitution of communication, so to speak, in modern societies consists of different communication systems — the systems that produce and transmit communication within an institutional framework. They engender communication by establishing a network of persons, techniques, institutions, standards, and codes, which together determine not only the practical conditions of communication but also to a great extent its conceptual form. I maintain that communication is always mediated along lines that cannot be completely contained within some pure types and controlled in an entirely rational manner, and that in contemporary societies these mediating instances have risen to a pivotal position. It is therefore of the highest importance to identify the forms in and through which organization of communication has been exercised successfully and that have spread across society and become generalized social techniques. Some of the most influential loci of the creation of new forms of knowledge relating to communication can be found in some confined institutional contexts, typically

5 True, insofar as it is considered of constituting human community, communication is thought of primarily inside philosophical tradition. However, the most influential tradition of thinking about modern communication is not found in philosophy, but rather in those fields in which communication has been seen as a technique.

connected to industrial or governmental practices. Generally speaking, these can be conceived of as the contexts in which communication has been given a systemic character and in which the project to rationalize communicative relationships was inaugurated.

It is precisely through these contexts of communication that we can approach the modern history of communication seen as a distinct theoretical and political subject. Once the first forms of telecommunication were established in about the middle of the nineteenth century, "communication" was made both an object and a means of governmental practices within a series of different discourses ranging from the theories of management to the suggestions for the reorganization of municipal governance. It was linked not only to regional organizational schemes but also increasingly to plans on a national scale, based on its integrating functions. What is more, communication by its systemic nature became a presupposition for thinking about the political community as an intelligible whole that could be subordinated to professional management. However, in order to be practicable, communication had to be formalized in a consistent theoretical framework, which made it an object of ever-intensifying investigation. This formalization was not restricted to a few specific communication systems alone but characterized the general attitude toward communication during the modern period, being fostered by the institutionalization of communication research. In this sense, the various systems of communication and related discourses and metaphors have invariably engendered frameworks within which communication has become an object of systematic thought and problematization — that is, it has come to be seen as a problem.

In this sense, communication always takes its place in some location or setting, which organizes it both materially and conceptually, although the perpetual realization of communicative actions generates new arrangements and forms of communication. These forms seem to be natural candidates for the objects of "pure sociology," proposed by Georg Simmel; namely, the "societal forms themselves." He distinguished the "pure" element of sociation from its heterogeneous and variable embodiments.[6] What Simmel had in mind were objective forms with positive contents rather than different communicative forms, although he ultimately failed to formulate a satisfactory theory of this kind of interaction. The fact that there seemed to be no *model* for a form that comes into existence through communication has been regarded since Simmel simply as a challenge. For instance, communication theoretician Hugh Duncan tried to draw up a model in 1962 that would explain a reciprocal relationship in which the forms of sociation are determined through communication, which, in turn, is constituted by these same forms.[7] The type of approach endorsed by Duncan and many others, which seeks to construct a positive model for communication, fails to take into account that lack of such a model is perhaps not due to a failing of theory, but instead a permanent state of affairs that any analysis of communication has to take into

6 See Georg Simmel, *The Sociology of Georg Simmel*, ed. Kurt Wolff (Glencoe, IL: Free Press, 1950).
7 See Hugh Duncan, *Communication and Social Order* (New York: Bedminster Press, 1962).

account. Now, by examining the diverse communicative assemblages as points of intersection of diverse technical, political, theoretical, and juridical practices, we have the possibility of considering communication in a Simmelian spirit as something pivotal to social life, but without Simmel's (and Duncan's) positivistic frame of analysis. This is because the systems in question do not passively reflect the functionality of social practices, but rather create fields of knowledge and practice in which this functionality is exposed to systematic re-reading and reorganized along the lines forged in this process.

The fact that communication lacks a fixed identity becomes clear from the assemblages of communication studied here: one can see that telegraphic communication has a different ontological constitution from that of telephonic communication — which is not so much a comment on their mutual technological differences as, more important, on the distinct realms of action they make possible. Hence, communication does not have the conditions or principles that would be valid in all communicational activity: it does not have such constancy in time that would allow it to be viewed as the same. The conception of communication is created over and over again within a system of special conditions of a given era. Yet on the other hand, in order to talk about the history of "communication" in the first place, some continuity has to be discernible, some sort of succession — not necessarily a coherent line of development but rather a set of practices and conceptions sharing the same *ethos*. Communication, here, is a multiplicity of heterogeneous series and fields, and this multiplicity is irreducible into any single identity in principle. In spite of this, the communicative constellations and their displacements traced here are significant since they have without doubt shaped the thinking of communication in a way that surpasses the scope of their specific contexts of origin, thus enabling certain continuity in the modern configuration of the idea of communication.

The history of communication presented here concentrates on telecommunications, meaning that the history of mass communication is omitted from the study and yet this is an integral part of the Western history of communication. Communication systems, in discussion about their increasing influence and social significance, have been addressed in the literature mainly in terms of the mass media, ultimately at the expense of the theories, practices, and institutions of telecommunications. How can the communications systems investigated here, which transmit primarily interpersonal communication, play a role in the articulation of the social significance of communication? This type of question is grounded on the conventional line of demarcation between public and private, which I think is not very helpful from the viewpoint taken here. Even private communication is always socially mediated and structured. What is more important instead is to analyze the forms and mechanisms of the social organization of communication. It is here that interpersonal communication technology, with the conceptual tools and the rhetoric relating to these technologies, provides the essential means of organization. Whereas the significance of the mass media is connected to wide dissemination of similar meaning-contents, intercommunication has to do with the functionality of social institutions, constituting the precondition for modern business organizations, industrial establishments, and governmental institutions.[8] What is central is the way the social is constructed and institutionalized *in* intercommunication.

The daily evolution of modern societies cannot be separated from this type of communication: it is not only a question of producing communication but also of producing *by* communication. Unlike in the case of the mass media, intercommunication is related largely to routines, to the accomplishment of daily tasks determined by the requirements of organizational objectives. In those routines, insofar as they serve some preset ends, the content of communication is usually formally subordinate to its functionality, and consequently the execution of an action is more significant than the individual meaning-contents transmitted. These routines reorganize social existence through constituting collective practices, rather than by interpretive representations in the first place. Thus, while forming the basic structure of the transmission of communication and information, intercommunication systems are a necessary condition of the daily functioning of modern society. In modern societies from the middle of the nineteenth century onward, the need to have organized communication networks has increasingly meant the strengthening of the position of telecommunication systems. They occupy a pivotal position as the precondition for thinking about these societies. In fact, it is only since the nineteenth century that they have emerged as *the* mechanisms for manifesting what is common: interlinked with all the other technological and cultural institutions that together make up society, they reveal the modern forms of being in common. While setting the frameworks for communication, they also execute it, and in the process of this communication its very point of departure, the context of its birth, is revised and transformed. It is this relationship that determines the constitutive status of communication within the actual operation of communicative practices in their social functions.

Second, it should be noted that my intention is not to study the "effects" of these technologies and communicative ideas on American society, because the mutual effective relationships between discursive and material processes cannot be reduced to any simple causal relationship. Given that the constitutive horizon of communication cannot be "external" to its actual happening as it is always already part of its daily realization, clearly the juxtaposition between "discursive" conditions and "real" ones cannot be regarded as being valid. Communication, here, is not a pure, privileged, and primary instance that would stand in contrast to something derived, artificial, and only apparent. It is determined and given its content in relation to a social set-up alone, only within which it becomes conceivable. Yet, this relationship is not unilateral because communication is the very condition of the self-determination and incessant reorganization of the constellation within which it is practiced. So the theories, metaphors, and discourses germane to communication establish such a discursive field in and through which

8 As for the telephone, see Ithiel de Sola Pool, *Forecasting the Telephone: A Retrospective Technology Assessment* (Norwood: Ablex 1983).

the construction of physical information and control systems could be motivated and advanced. On the other hand, once becoming established these systems, in turn, supported and strengthened the underlying ideas and conceptions.[9] In this way, it is generally speaking impossible, in a given historical period, to distinguish clearly techniques, institutions, theories, and the metaphoric resources from one another and to consider each as completely independent, because they become intertwined and receive justification for their existence from one another. In this sense, the focus of the book is on what can be called the material ontology of communication: it studies not the meanings of communication as such but rather the material conditions and structures that allow those meanings to be formulated in the first place.

Let me now, lastly, introduce the structure of the book after this opening chapter. In the second chapter I deliberate on how the communicative sphere in the last half of the nineteenth century, seen from the viewpoint of telecommunication, came to be structured mainly as a result of the conjunction of two separate force fields. These were characterized by a concern for political unity, on the one hand, and by the advent of telegraph technology, on the other hand. Once interwoven, the two fields formed a new communicative order. Telegraph lines, or "lightning lines" as they came to be known, provided one of the most powerful models for representing society. I make the case that the telegraph encapsulated the conception of the technical basis of political unity. The chapter examines the way in which the development of transportation and the communications systems played a part in bringing coherence to a national community. This is followed in the second part of the chapter by an examination of the role of the telegraph in creating centralized communication structures in the same manner as the nervous system functions in a living organism.

The ideas concerning communication underwent a change at the beginning of the twentieth century. This was mainly because of the interweaving of two distinct fields of power: first, rational planning and scientific management, and second, the technology of telephony. Together these fields came to draw up a novel diagram of communication. The third chapter investigates each theme in turn. First, I examine the theoretical tradition from Lester Ward through Frederick Taylor to the technocracy movement that formed the conceptual terrain for thinking about communication during this period, after which I turn to the forms and modes of communication as realized in certain telephone-based systems. Particularly, I elaborate on the component position of the telephone system with respect to the growth of hierarchically controlled organizations. I show how biological analogies were gradually replaced by mechanical ones, or at least were interpreted in light of them.

The fourth chapter investigates the emergence of a new form of communicative order after World War II. This new order came into being as a result of the

9 For the cybernetic language in particular, see Paul Edwards, *The Closed World: Computers and the Politics of Discourse in Cold War America* (Cambridge: MIT Press, 1996), 99.

interweaving of two processes. First, the systems-scientific approach spread and became common; second, information and communication technology developed and came to be seen as a necessity. Together these two processes engendered a fairly consistent form of communication that did not begin to weaken until the 1970s. As a result, the totality within which communication and central social processes were conceptualized changed from "machine" to "information system." In this chapter, I examine the practices within which communication was redefined as a distinct subject and means of governance in the computer age. In particular, I focus on how the concept of an automatic system began to steer thoughts about both communication and society during the Cold War period. The first part of the chapter will investigate the concepts of communication based on cybernetics and information theory, whereas the second part is devoted to the relationships these ideas opened up in thinking about "the social" in general.

In the concluding chapter, in which I bring together the history of a "modern" conception of communication, I argue that it is the very metaphor of the nervous system that has served as the central conceptual framework for understanding communication throughout the age of telecommunications. This metaphor gave a name to the model that brought a certain unity to the different mechanisms and constellations of communication. Although the nature of this model did not remain the same but evolved according to changes in and of communication from an organic to a feedback structure, there is a certain consistency underlying the vicissitudes of communication.

I would like to thank the London School of Economics and Political Science, the British Library, the Academy of Finland, the University of Helsinki, and the Finnish Cultural Foundation for the invaluable help they provided in terms of funding, expertise, or locating some of the sources needed.

2

Lightning Lines

The communicative sphere in the last half of the nineteenth century, seen from the viewpoint of telecommunication, came to be structured mainly as a result of the conjunction of two separate fields of forces. They were characterized largely by a concern for political unity, on one hand, and the advent of telegraph technology, on the other hand. Once interweaved, they formed a new communicative order that replaced a system of local and regional networks by the late nineteenth century. Telegraph lines, or "lightning lines" as they commonly came to be known, started to spread vigorously under private control in the middle of the nineteenth century, and came to represent not only a tool for political integration but also a resource for popular images and societal metaphors.[1]

When the telegraph came to characterize a whole era by providing one of the most powerful models for the representation of society, modernity differentiated itself from its historical past. In the eyes of the moderns, the telegraph encapsulated something that had never before found historical expression in the realm of communication. If this experience of a new era could be crystallized, it would, perhaps, be the conception of the technical basis of the political community. In other words, what was at issue was the fact that communication as a technique — telecommunication — seemed to foster the community as a political entity at a time when disintegration was a heavy threat to the early republic. What was conceived of in terms of unity referred not only to the republic as a political system but also, potentially, to a spiritual community coming between independent nations. Here, communication was held to constitute the preconditions for a truly political (national) and spiritual (pan-national) unity to be engendered.

The development of telecommunications has been essentially connected with the founding of a community insofar as the latter has been understood as a social

1 For early telegraph manuals and histories, see Alfred Vail, *The American Electro Magnetic Telegraph with the Reports of Congress, and a Description of all Telegraphs Known, Employing Electricity or Galvanism* (Philadelphia: Lea & Blanchard, 1845); Charles Archer, *Guide to the Electric Telegraph* (London: Smith, 1852); Edward Highton, *The Electric Telegraph: Its History and Progress* (London: John Weale, 1852); Alexander Jones, *Historical Sketch of the Electric Telegraph: Including Its Rise and Progress in the United States* (New York: George P. Putnam, 1852); Laurence Turnbull, *Lectures on the Electro-Magnetic Telegraph* (Philadelphia: R. W. Barnard, 1852); Charles Briggs and Augustus Maverick, *The Story of the Telegraph, and a History of the Great Atlantic Cable* (New York: Rudd & Carleton, 1858); George Wilson, *The Progress of the Telegraph* (Cambridge: Macmillan, 1859); George Prescott, *History, Theory, and Practice of the Electric Telegraph* (Boston: Ticknor & Fields, 1860); Dr. Lardner and Edward Bright, *The Electric Telegraph* (London: James Walton, 1867).

project. Seen from a technological viewpoint, it has, in the first place, invariably been regarded in America as having determined the technical basis for the promotion of political unity.[2] On the other hand, by engendering more or less common national publicity, telecommunication systems — with the telegraph leading the way — also created the network of conditions and possibilities necessary for big-business organization — the principal user and developer of this mode of communication. This is why the history of modern communicativity — that is, the shape or image of the possibility of communication — is closely tied to development of the conditions of commercial activity and to the concentration of economic power. This constituted a context for thought about communication; in other words a context in which communication could be brought up as a distinct object of thought. To an ever-greater extent, however — in the form of improving communication systems — it constituted the environment for the emerging social and economic order.

Let me begin by examining the way in which the development of transportation and communications systems has played a part in bringing coherence to a national community. This is followed, in the second part of this chapter, by an examination of the telegraph's role in the creation of centralized communication structures.

Coming Unity

The introduction of the commercial telegraph in 1844 could be seen as having been an event without a present tense. This is true in two senses. First, the telegraph, conveying "intelligence to any given distance with the velocity of lightning,"[3] inaugurated a new era of electric communications that lacked an obvious institutional and technological parentage.[4] With the improving transportation systems, it destroyed the traditional order of life and gradually made it necessary to erect institutional frameworks for its social control. The telegraph was thought about through existing social institutions and practices that provided indispensable yet insufficient cultural contexts for it. Because of its uniqueness, the telegraph demanded, according to communications theorist James Carey, "a new body of law, economic theory, political arrangements, management techniques, organizational structures, and scientific rationales."[5] It was thus a novel social form without any real-life equivalent. Second, and more generally, any new event, it can be said, has no present. Rather, it retreats and advances in two directions at once; namely, to the past and to the future. Both of these form complete and distinct realms in themselves, each dividing any singular event into the past and the future. A new

2 This is not to say, of course, that telecommunication constituted the only or the privileged basis of the political community, but, rather, that it was an important constituent of this basis. On technology seen as the means with which to preserve the Union and to promote a sense of unity in the early republic, see John Kasson, *Civilizing the Machine: Technology and Republican Values in America, 1776–1900* (Harmondsworth: Penguin Books, 1977).

3 Turnbull, *Electro-Magnetic Telegraph*, 3.

4 Richard Schwarzlose, *The Nation's Newsbrokers: Vol. 1, The Formative Years, from Pretelegraph to 1865* (Evanston: Northwestern University Press, 1989), 41.

5 James Carey, "Technology and Ideology: The Case of the Telegraph," in *Communication as Culture: Essays on Media and Society*, ed. James Carey (London: Unwin Hyman, 1989), 205.

event is always determined by these two realms; it is a borderline case within the scope of a given temporal direction that it reflects and expresses.[6]

The advent of the telegraph could be understood through this interpretation as being linked with both the past and the future. On one hand, the telegraph was conceived in accordance with general aims that the existing arrangement for information distribution — the postal system — served, especially with that of integrating the nation by disseminating public information. On the other hand, when its unifying potential was understood, the telegraph soon captured public imagination as the bond enabling future peace and harmony among different peoples and nations. These were not two separate phases of the development of the telegraph, but instead closely interrelated ones, centering on the telegraph's unifying capabilities. In both cases there was the same underlying logic that reflected the contemporary conception of the nature of this first form of telecommunication: it was conceived of as a fundamentally neutral instrument of communication, the benefit of which was to establish connections within society without governmental interference. That conception made it possible to formulate politically agreeable projects for national unification in terms of communication and to universalize the hopes and plans invested in those. Let me now elaborate on these themes a little further.

Wiring the Nation

As telegraphic communication was introduced in the United States, it was incorporated into the general political context in which the postal system figured centrally. The American postal system was as a focal governmental organization that linked the demands of communications and governance. Governance was needed, because nineteenth-century America was a loose association of poorly connected settlements, "island communities," to use Robert Wiebe's well-known expression.[7] In this dispersed societal structure, based on the principle of local autonomy, weak communications rendered interregional interaction difficult, thus impeding public participation in politics and enactment of public policy. The problem of communication was therefore one of the most important ones national government had to solve, being as much a technical as a political question.[8] There was a genuine concern regarding the limits to extension of the Union as an integrated political unit in a period of rapid westward movement coupled with fragmentary and unpredictable communication and transportation systems.[9] Thomas Ewing, Secretary of the Interior, urged "some means of communication

6 See Gilles Deleuze, *The Logic of Sense*, trans. Mark Lester with Charles Stivale, ed. Constantin Boundas (London: Athlone Press, 1990), 80.

7 Robert Wiebe, *The Search of Order, 1877–1920* (London: Macmillan, 1967), xiii.

8 Wiebe, *Search of Order*, xiii; see also Robert Russel, *Improvement of Communication with the Pacific Coast as an Issue in American Politics, 1783–1864* (Cedar Rapids: Torch Press, 1948); Nelson Klose, *A Concise Study Guide to the American Frontier* (Lincoln: University of Nebraska Press, 1964); James Nagle, *A History of Government Contracting* (Washington, DC: George Washington University, 1992).

9 The question of whether there are some natural limits to democracy beyond which it would lose its viability as a system, discussed already by James Madison in the Federalist Papers, contributed to the suspicion toward the extension of the republic (see paper number 14 in Alexander Hamilton et al., *The Federalist*

across the continent, through our own territory, from the Atlantic to the Pacific" in 1849, thus expressing the need to link the two coasts.[10] In the following year President Millard Fillmore brought up the same question — as Zachary Taylor had done earlier in his annual message as president.[11] It was crucial to link settlements by reliable and effective means to the main body of the Union in order to maintain political order and coordinate the scattered system of regions into the framework of one political system. The postal system was the foremost national institution developed in accordance with the objective to "facilitate the administration of Government: and more especially, in a country like ours, to make known, by a rapid circulation, the political disquisitions relative to public measures," as the matter was put in "the Report upon a National Road, Washington to New Orleans," becoming the central administrative apparatus of the nation by the early nineteenth century.[12]

There were at least two reasons to further effective and reliable dissemination of political information. First, according to a republican doctrine that was an essential part of the theory of government from Thomas Jefferson's sponsorship of a "Bill for the More General Diffusion of Knowledge" of 1778 onward, an informed and vigilant public constituted the foundation of liberal governance, forming an ultimate defense against a strong government. Participation of a politically aware citizenry in the affairs of national politics thus became an essential republican principle.[13] Second, a need to Americanize the growing number of immigrants forced the administration to seek means of developing a comprehensive system for information distribution.[14] These policies were conducive to growing a comprehensive communication and transportation network, which became the technical basis for an evolving structure of information distribution.

The postal system was a tool under the government's control for binding localities together, for decreasing the isolation of settlements, and for tightening the mutual interdependence of the regions, thereby enabling the feeling of a common national space in both political (a truly federal unit) and economic

Papers (New York: New American Library, 1961), 99–105; see also Russel, *Improvement of Communication*, 10–11).

10 "Report of the Secretary of the Interior," House Ex. Doc. 5, 31st Cong., 1st Sess. (1849).

11 "Message from the President of the United States to the two houses of Congress," Senate Ex. Doc. 1, 31st Cong., 2d Sess. (1850).

12 House Doc. 156, 19th Cong. 1st Sess., (1826), 24; Richard John, *Spreading the News: The American Postal System from Franklin to Morse* (Cambridge, MA: Harvard University Press, 1995), 25.

13 Frank Tariello Jr., *The Reconstruction of American Political Ideology, 1865–1917* (Charlottesville: University Press of Virginia, 1982), 29; Russell Hanson, *The Democratic Imagination in America: Conversations with Our Past* (Princeton: Princeton University Press, 1985), 76; Richard Brown, *Knowledge is Power: The Diffusion of Information in Early America, 1700–1865* (New York: Oxford University Press, 1989), 292–3. As the cornerstone of its postal policy, Congress speeded up the diffusion of knowledge by subsidizing newspaper postage, thus extending the authority of the government, but also, and important, providing the citizenry with the means to monitor its elected representatives (Wayne Fuller, *The American Mail: Enlarger of the Common Life* (Chicago: University of Chicago Press, 1972), 113–47; John, *Spreading the News*, 31–63). Newspapers remained the main means of publishing laws and other official information of political importance until the 1870s (Culver Smith, *The Press, Politics, and Patronage: The American Government's Use of Newspapers, 1789–1875* (Athens: University of Georgia Press, 1977), 245–6).

14 Fuller, *American Mail*, 134.

(a national market) terms. In fact, it seemed possible, as a commentator put it in 1829, that unlike any other mechanism of intercourse, the postal system could transform the whole country from a confederation of separate states into "one great neighborhood,"[15] thus anticipating Samuel F. B. Morse's similar description of the power of the telegraph. As historian Richard John pointed out, long before the telegraph was credited with having "annihilated" space and time, the postal system had been described in exactly the same terms.[16]

So, the concept of unification — an idea that did not always get explicitly formulated but became rather a *de facto* basis for thinking about the nation as one political entity — implied in its technical structure two interlinked political tasks. First, it introduced the requirement to create physical communication and transportation connections. This was the dimension of roads, embankments, locks, intersections, and tracks constituting a network of trails over the surface of a wide terrain. Second, there was a need to construct a comprehensive and effective distribution system for public information. This latter task naturally relied on the existence of an adequate physical infrastructure, but it had a distinct dynamic of its own. More important than to coordinate the evolving set of tracks was to create an arrangement for securing easy circulation of different types of information over distance.

If both of these tasks were necessary for bringing about a consistent and harmonious political community, the objective in itself could be seen as having two distinct yet related parts; namely, political stabilization and cultural homogenization. Political stabilization accounted for governmental attempts to mold the country into a coherent political unit with its central political institutions and legislative frameworks covering the whole country in an equal fashion. The need for cultural homogenization, for its part, showed itself in the emphasis on the common national ground of the American people, irrespective of differences in ethnic background, geographic location, or individual interests.

In this way, all these projects were centered on the same goal; namely, rendering the nation more like one political and economic unit. The postal system provided the vehicle to bring these objectives together in a way that allowed not only governance of the population but also counter-governance of the government by the people, thus "drawing still closer the bond of union," as a congressional committee put it in 1830.[17] However, the urge for yet speedier communication had induced Postmaster General John McLean to consider the introduction of a telegraph system for postal use already in 1828. At that time the term "telegraph" referred to the semaphore system, invented by a French engineer called Claude Chappe, of transmitting visible signals among towers located about ten miles from one another in a chain.[18] Although the system was widely copied by other

15 William Ellery Channing, "The Union," *Christian Examiner* 6 (1829). Quoted in John, *Spreading the News*, 13.

16 John, *Spreading the News*, 10.

17 Quoted in Fuller, *American Mail*, 81.

18 Patrice Flichy, *Dynamics of Modern Communication: The Shaping and Impact of New Communication Technologies*, trans. Liz Libbrecht (London: Sage, 1995), 8–13, 16–18, 29–32.

European countries, Congress never authorized the construction of this kind of optical telegraph, even though pressure to restrain growing market speculation by some effective means of communication — faster than any private express service in delivering market information — was imperative. It seemed that the electric telegraph, the "last and most wondrous birth of this wonder-teeming age,"[19] would suit this purpose perfectly.

Insofar as there was, at the introduction of the telegraph, any feasible model of the politicization of communication — in other words of communication understood as both the object and the means of political projects — then it must have been closely linked to the postal system.[20] This can clearly be seen in political discussions concerning at least two kinds of questions: first, whether or not the telegraph should be under the control of the post office so as to be used in the same way as mail to give news to everyone at the same time, and second whether telegrams were similar to or analogous with letters, as an applicable legislative framework in as far as the telegraph was needed.[21] In both cases the postal system seemed to provide a natural model for thinking about the new questions and issues the telegraph evoked, which were not limited to ownership and legislature but were related to wider problems concerning nation-building.

Thus it was the postal system again — including all the political functions and imagery invested in it — that established the framework through which the telegraph appeared as a critical means, a key political technology, to be harnessed in building up national cohesion. Regarding the complex processes of territorial expansion and population increase with their unanticipated concomitants in the society of mid-nineteenth-century America, the continental extension of communication and transportation systems became crucial in maintaining political consistency: the telegraph, coupled with the railroad, became pivotal technology of social control. Besides easing information diffusion, the telegraph, and the railroad,

19 House Report 187, 28th Cong., 2d Sess. (1845), 3.

20 Telecommunication, traditionally understood, was not politicized until the end of the nineteenth century, when it became a political tool in the rivalry among the great powers (e.g., Daniel Headrick, *The Invisible Weapon: Telecommunications and International Politics, 1851–1945* (New York: Oxford University Press, 1991), 7; Daniel Headrick, "Public–Private Relations in International Telecommunications Before World War II," in *Telecommunications Politics: Ownership and Control of the Information Highway in Developing Countries*, ed. Bella Mody, Johannes M. Bauer, and Joseph D. Straubhaar (Mahwah: Lawrence Erlbaum Associates, 1995), 37–8). However, here the term refers to political interest in the telegraph as a strategic apparatus for consolidating the nation: this made it an essential political technique prior to its role in international rivalry.

21 Fuller, *American Mail*, 172–4; Lester Lindley, *The Constitution Faces Technology: The Relationship of the National Government to the Telegraph, 1866–1884* (New York: Arno Press, 1975), 45–9. According to Postmaster General Howe, writing in 1882, "the business of the telegraph is inherently the same as that of the mail. It is to transmit messages from one person to another" (cited in Katharine Judson, ed., *Selected Articles on Government Ownership of Telegraph and Telephone* (White Plains: H. W. Wilson, 1914), 89). Shipping goods by common carriers was also presented as analogous to transmitting messages by wire, and was thus seen as providing reasonable precedents for telegraph law (e.g., *Telegrapher*, 1, (2), (October 31, 1864): 9, 15). By analogizing from the law of carriers, lawyers attempted to adjust the existing legislative framework to the new circumstances of an industrial society; see Lester Lindley, *The Impact of the Telegraph on Contract Law* (New York: Garland, 1990), 30–70, 125–9, 137.

facilitated communication to remote forts and settlements, thereby increasing capabilities to improve military defense and coordination, and rendering "our system of Government easily and certainly maintainable."[22] The growing network of connections was thus conducive to increasing safety, predictability, and homogeneity.

With the introduction of the telegraph, speed came to be established as a new factor that would rearrange the conditions of communication and set this technology ultimately apart from the postal system. The telegraph was fast: "[information] can be almost instantaneously transmitted between any two or more points," said Samuel Morse explaining his new invention.[23] This speed encouraged people fascinated by this communication facility to draw on the ideas associated with the postal system — ideas that were subsequently repeated in connection with every new communication technology — and to claim that "time and space are annihilated."[24] The telegraph differentiated flows of information according to factual value, thereby dividing messages into different classes and types that had previously seemed to constitute a rather uniform group: "The first result of the discovery [of electricity] has been a vast diffusion of what is called 'news,' the recording of every event, and especially of every crime, everywhere without perceptible interval of time."[25] Telegraphic communication was fundamentally about transmitting messages with a particular pecuniary or news value. News-gathering was thus transformed into a dynamic process and the status of news was based strictly on novelty.[26] Information in its sphere moved rapidly, whereas other contemporary literary or journalistic materials continued to circulate relatively unhurriedly. More important, these rapidly transmitted telegrams now became the norm: they defined the type of information that was considered to be of particular social importance.

As a result, the telegraph introduced a completely new idea of communication. Now for the first time communication could potentially become a continuous flow supplanting the preceding forms of intermittent waves of information supply. It was the telegraph that made communication a continuous process, thus freeing it from the rhythmical motion of the Postal Service and from physical transportation in general. The advent of the telegraph did not, of course, take the place of the postal system as a mechanism for general information delivery, but what it did do was establish a brand new conception of communication, differentiating intermittent torrents from a continuous flow, and, for the first time, transportation

22 Postmaster General Mr. Johnson in a report by the Ways and Means Committee of the House of Representatives, published in 1845; cited in Congressional Record, 48th Cong., 1st Sess., vol. 15 (1884), 376. See also Jeffrey Kieve, *The Electric Telegraph: A Social and Economic History* (Newton Abbot: David & Charles, 1973), 241.

23 Morse's answer to Levi Woodbury, the Secretary of the Treasury, on September 27, 1837, in Vail, *American Electro Magnetic Telegraph*, 70. By 1873 the telegraph was capable of a speed of between 500 and 800 words per minute, as opposed to the average speed of fewer than 25 words by an expert Morse operator (*Annual Report of the Postmaster General: The Fiscal Year Ended June 30, 1873*, 45).

24 Marshall Lefferts, "The Electric Telegraph: Its Influence and Geographical Distribution," *Bulletin of American Geographical and Statistical Society* 2 (January 1857): 262.

25 "The Intellectual Effects of Electricity," *Spectator* 63 (November 9, 1889).

26 Frank Mott, *The News in America* (Cambridge, MA: Harvard University Press, 1952), 42–3; Menahem Blondheim, *News over the Wires: The Telegraph and the Flow of Public Information in America, 1844–1897* (Cambridge, MA: Harvard University Press, 1994), 56.

from communication.[27] This was the moment that marked the beginning of the course of action in which information came to be utilized as a means for controlling other technological processes, especially in the railroads, and in which the telegraph emerged as both a model of and a mechanism for this control.[28] In this way, by providing speed, regularity and reliability, it started to play an important part in creating and maintaining areas of stability.

By liberating information transfer from transportation, the telegraph expanded the boundaries of a region within which the regular exchange of information could be routinely maintained.[29] This required immense investments in the construction of a comprehensive system of communication lines. Making the West Coast reachable from the East Coast by telegraphic means was a major development in the creation and extension of areas of stability. The military, commercial, and administrative value of the line to California was clear to the government in that it decided to subsidize its construction, as stated in the Pacific Telegraph Act of 1860. It thereby hoped to bring into closer interconnection not only the ends of the line — with specific reference to the western coastline — but also the broad areas in between.[30] The major political reason for the need of telegraph connection

27 This remark has been made many different times in numerous connections (e.g., in Arthur Page, "Social Aspects of Communication Development," *Bell Telephone Quarterly* 11, (2), (1932): 167; Arthur Page, "Social Aspects of Communication Development," in Arthur Page et al., *Modern Communication* (Boston: Houghton Mifflin, 1932), 8). In the social sciences it was articulated perhaps most famously by James Carey; see "Technology and Ideology"; "Time, Space, and the Telegraph" in *Communication in History: Technology, Culture, Society*, ed. David Crowley and Paul Heyer (New York: Longman, 1991). For a commentary of Carey's views, see Richard John, "American Historians and the Concept of the Communications Revolution," in *Information Acumen: The Understanding and Use of Knowledge in Modern Business*, ed. Lisa Bud-Frierman (London: Routledge, 1994), 106–8.

28 James Reid, *The Telegraph in America: Its Founders, Promoters and Noted Men* (New York: Derby Brothers, 1879), 480; Robert Thompson, *Wiring a Continent: The History of the Telegraph Industry in the United States, 1832-1866* (Princeton: Princeton University Press, 1947), 443. George Taylor, *The Transportation Revolution, 1815-1860* (Rinehart & Co., New York, 1951) is a classic account of the "transportation revolution": for a more recent interpretation that seeks to break away from the habit of explaining the rise of the railroad in terms of the dynamics of efficiency, see William Roy, *Socializing Capital: The Rise of the Large Industrial Corporation in America* (Princeton: Princeton University Press, 1997), 78–114. For a discussion on the telegraph's influence on long-term economic growth, especially in the American railroad sector, see Alexander Field, "The Magnetic Telegraph, Price and Quantity Data, and the New Management of Capital," *Journal of Economic History* 52, (2), (1992).

29 Irwin Lebow, *Information Highways and Byways: From the Telegraph to the 21st Century* (New York: IEEE Press, 1995), 15.

30 As regards the military aspect, Mr. Colfax, for instance, stated in 1860 that "in the case of war, we should have a communication with the Pacific of the speediest possible character; for should we become embroiled in war with a foreign nation [. . .] it is of the utmost importance that such a means of [telegraphic] communication should be established, for upon that Pacific region would the heaviest blow fall; because there is a long line of exposed coast, and our cities and ports upon that coast are without fortifications, and comparatively without defense" (Congressional Globe, 36th Cong., 1st Sess, 1860, [1693]).

See also Reid, *Telegraph in America*, 414; Walter Towers, *Masters of Space* (New York: Harper & Brothers, 1917), 97; Alwin Harlow, *Old Wires and New Waves: The History of the Telegraph, Telephone, and Wireless* (New York: D. Appleton-Century Co., 1936), 306–11; Harold Sharlin, *The Making of the Electrical Age: From the Telegraph to Automation* (London: Abelard-Schuman, 1963), 31–2; Phil Ault, *Wires West* (New York: Dodd, Mead & Co., 1974), 44–5, 74–6; Paul Israel, *From Machine Shop to Industrial Laboratory: Telegraph and the Changing Context of American Invention, 1830-1920* (Baltimore: Johns Hopkins University Press, 1992), 55; Edwin Gabler, *The American Telegrapher: A Social History, 1860-1900* (New Brunswick: Rutgers University Press, 1988), 39.

was the fear that rising cities in the far west could disintegrate the Union and make its separation necessary.[31] The words of President Abraham Lincoln through the new line, which was completed in the following year, eight years prior to the transcontinental railroad, besides showing great relief, were indicative of the belief in the harmonizing effects of this new connection to "the stability and union of the Republic."[32] This conviction seemed increasingly to be justified as the telegraph began bringing the various self-sufficient "island communities" into more intimate contact with one another by means of a constant flow of information.[33] The worry about disintegration appeared to lose some of its urgency at an age when the Americans, according to Mr. Yates of Illinois, were wont to "travel by steam, print by electricity, and talk by lightning."[34]

Now, it was evident that the telegraph, coupled with the railroad, played an essential part in plans for national unification and homogenization, an idea that had strengthened by the time of the threat of, and continued to grow stronger especially after, the Civil War.[35] Therefore a legislative framework was needed to create a predictable environment for favorable extension and consolidation of the nascent communications network. The Act that Congress passed in 1866, granting telegraph companies the right to build lines "over and along any of the military or post roads of the United States," served as a basis for not only organizing a rudimentary form of regulation but also a way of supporting companies in their efforts to spread telegraph lines across the country.[36]

As the establishment of economic links between different regions was widely considered the best way of bringing the nation together, the extension of the telegraph lines in addition to the development of the railroad system seemed to be perfect tools toward that end. For as the then Secretary of State William H. Seward stated: "Other conditions being equal, the country that has the largest extension and the most thorough radiation of the telegraph wire enjoys the most active and profitable system of domestic commerce."[37] In effect, after a period of suspicion

31 US Secretary of State William Seward, for instance, expressed such concern; see Paul Nagel, *The Sacred Trust: American Nationality 1798–1898* (New York: Oxford University Press, 1971), 55. The communication question was also entangled with broader political problems such as the slave question, as many people assumed that, although California had come to the Union formally as a free state in 1850, without telegraphic connection it could, it was feared, become politically bound to the South (see Russel, *Improvement of Communication*, 25).

32 Quoted in Ault, *Wires West*, 74.

33 Blondheim, *News*, 194.

34 Appendix, Congressional Globe, 32d Cong., 1st Sess., (April 23, 1852), 476 (cited in D. W. Meinig, *The Shaping of America: A Geographical Perspective on 500 Years of History; Vol. 2, Continental America, 1800–1867* (New Haven: Yale University Press, 1993), 348).

35 See Alan Marcus and Howard Segal, *Technology in America: A Brief History* (San Diego: Harcourt Brace Jovanovich, 1989), 90; Morton Keller, *Affairs of State: Public Life in Late Nineteenth Century America* (Cambridge, MA: Belknap Press of Harvard University Press, 1977), 161, 167.

36 *Telegraph Lines* (July 24, 1866) (quoted in United States of America, War Department, Document No. 104, *Regulations for the Operation and Maintenance of United States Military Telegraph Lines and General Regulations of the Signal Corps*, United States Army, Government Printing Office, Washington, DC, 1899). See also Tal Shaffner, *The Telegraph Manual* (New York: Pudney & Russell, 1859), 763, 764, 766; Harlow, *Old Wires*, 331.

37 William Seward, *Communication of Hon. William H. Seward, Secretary of State, upon the Subject*

about its supposed advantages, it gradually became clear that the telegraph would constitute a key asset in the modernization of the economy. "The function of the telegraph in our highly organized commercial and social life," pronounced *Harper's Weekly*, "has come to be as general and as important as that of the mail."[38] Yet the telegraph could transmit messages within a single day over a distance that by mail would have taken possibly weeks, becoming thus an unbeatable communications device whenever speed was required. As people became convinced of the practicability of the telegraph and as it started to spread across the country, usually following post roads and later railroads, separate communities could receive the same information almost simultaneously: now news of changes in amounts of cotton production in the South was transmitted almost instantly to every major financial center, from which the implications reverberated over the whole nation.

The telegraph thus appeared to be a powerful mechanism for fostering truly national markets by expanding the area within which goods could be bought and sold.[39] It radically reduced the time required to send market information across great expanses, thus facilitating the spread of a uniform price system that decreased what is called arbitrage — buying cheap and selling dear by transporting goods around — and the emergence of large, efficient markets. By putting everyone on an equal footing for trade purposes, it universalized and "decontextualized" the markets — in other words, removed them from the particular local context and placed them in an ever more extensive area of commercial activity that worked as a single structure.[40] Thus, "words became the tokens with which trade was carried out," in the words of a commentator in 1933.[41]

Significantly, the telegraph not only contributed to the extension of markets by

of an Intercontinental Telegraph Connecting the Eastern and Western Hemispheres by Way of Behring's Strait (Washington, DC: Government Printing Office, 1864), 17.

38 Editorial in *Harper's Weekly* 27 (July 28, 1883): 467.

39 J. C. Furnas, *The Americans: A Social History of the United States, 1587–1914* (New York: G. P. Putnam's Sons, 1969), 378; Richard Du Boff, "The Telegraph and the Structure of Markets in the United States, 1845–1890," *Research in Economic History* 8 (1983); Richard Du Boff, "The Telegraph in Nineteenth-Century America: Technology and Monopoly," *Comparative Studies in Society and History* 26 (1984): 571–86; JoAnne Yates, "The Telegraph's Effect on Nineteenth Century Markets and Firms," *Business and Economic History*, 2d Series 15 (1986); Blondheim, *News*, 172; for the so-called market revolution in general, see Melvyn Stokes and Stephen Conway, eds., *The Market Revolution in America: Social, Political, and Religious Expressions, 1800–1880* (Charlottesville: University Press of Virginia, 1996). The telegraph provided a wide variety of means for conducting business operations. F. W. Lienau, a superintendent of the Tariff Bureau of Western Union, for instance, said that "the telegram [. . .] is an effective medium [. . .] for placing business propositions before a number of people at once [. . .] Again, business men are developing the habit of using the telegram in keeping in touch with their field forces and their salesmen and encouraging their activities, in cultivating closer contact with their customers, in placing their orders, in replenishing their stocks, and in any number of other ways calculated to further the profitable conduct of their enterprises" (Cited in Towers, *Masters of Space*, 280). However, it has also been pointed out that notwithstanding the improvements in communications, it was the role of the Court that was pivotal in building a continent-wide open market (Archibald Cox, *The Court and the Constitution* (Boston: Houghton Mifflin, 1987), 101).

40 Carey, "Technology and Ideology," 217–22. Overall, the history of telecommunications in the late nineteenth century is said to have been a history of business more than politics (Headrick, *Invisible Weapon*, 6–7; Headrick, "Public–Private Relations," 37–8), although the creation of nationwide markets was clearly an overarching political task *par excellence*.

41 R. T. Barrett, "The Growth of America's Communication Needs," *Bell Telephone Quarterly* 12, 4 (1933): 267.

providing a fast communication system for sending market information but also made the very communication a product that could be sold on the market. This was obvious in the telegraph-dependent news-brokering reporting in which all stories were made marketable commodities, and it was also manifest in everyday telegraphy. It became a common characteristic of all communication systems that communication was something that could be sold: communication, and especially a communication event, was made a saleable commodity. All systems of communication yield and organize communication, but not until the commercial telegraph became common was it included in capitalist production. Before this, communication had mainly been a resource outside market relationships, but the telegraph made it an article of trade, in the form of a telegram, on which this communication system based its industrial organization. Thus, for the first time communication as an event was incorporated into the market economy and became an industry of its own.[42] It was these circumstances that politicized the communicative sphere of the telegraph.

Another reason for political interest in the telegraph was the aforementioned need to inform the public — a goal successfully advanced by the postal system. This was an important issue because it was this policy to keep the public informed about political matters that was regarded not only as a means of political consolidation but also as the very precondition of public governance. In addition, with the help of the telegraph it linked demands for local autonomy with the task of unifying the Union politically. It was in this context, in addition to its role of fostering the national marketplace, in which the telegraph assumed its most significant political function. Increasing social circulation of information by means of the telegraph was an attempt to diminish differences in availability of information between regions and localities. Once the telegraph assumed the character of a network, it became a powerful means for this particular purpose. Thus, systems were established by which news and governmental information could be sent simultaneously from a central station to a number of points, enabling reception of congressional reports, for example, at the same time in Baltimore, Philadelphia, and New York.[43] In this respect, the Wire Service that later became a part of the New York Associated Press grew into a central institution, transmitting uniform news to numerous newspapers over the country by utilizing precisely the network nature of telegraph lines.

Improved communication and transportation capabilities coupled with expanding political and news reporting by wire service advanced the forming of public opinion, which, in turn, was a quintessential element in a process of political

42 True, a price was also put on communication in the postal system, but it was not an event or an industry, at least not in the sense commonly understood today. Telegraphic communication, on the other hand, was both an event and an industry, although it was not interactive. Lester Lindley claimed that the telegraph was not an article of trade because it expedited commerce rather than restrained it, which is what the typical monopolies did (Lindley, *Constitution*). However, even though the telegraph was not a commodity, the telegram was. Elsewhere, Graham Murdock has argued that the rise of modern communications has been irrevocably bound up with the commodification process, which is incorporating more and more areas of communicative activity (Graham Murdock, "Communications and the Constitution of Modernity," *Media, Culture and Society* 15 (1993): 528).

43 Thompson, *Wiring a Continent*, 248.

integration.[44] The increasing public availability of information, dating back to the governmental policy of diffusing knowledge by subsidizing newspaper postage, gave rise to the idea of a common sphere that existed not in a particular place but rather in the imaginations of people, an idea similar to what has been described regarding the effects of the postal policy.[45] As James A. Garfield, then a member of Congress put it, "distance, estrangement, isolation, have been overcome by the recent amazing growth in the means of intercommunication. [. . .] Public affairs are now more public, and private less private, than in former ages."[46] If the telegraph had brought, in a certain crucial respect, transparency to the economic realm, it was hoped the same could take place in the political domain as well. Although it constituted only the latest improvement in already relatively rapid and reliable communications and transportation systems, the telegraph was soon established as a basis for modern systems of information distribution allowing individual views and interests to get in touch with one another. By itself becoming part of the social fabric, it redefined the conditions for a feeling of nationality, as, for example, the telegraph historian Robert Thompson observed when he concluded that "a stronger and more unified nation [. . .] emerged with the development of the telegraph and the railroad."[47] The telegraph thus constituted a means for national ends, but in addition to this, and more important, it was also a means for constitution of the political community as a national experience.

It is true that the telegraph and railroad — in conjunction with the extension of education and literacy — unified populations and made elements, sections, and regions more interdependent in late-nineteenth-century America, becoming a popular theme also in the literature of the period. An assemblage of practices, institutions, and technologies centering on the telegraph laid the foundations of a systematic and effective mechanism for channeling a continuous flow of information across the nation. However, the process of harmonizing the country was not a matter exclusively of the scope of information routes, of the nature of the material moving over and within these networks, or of the speed of individual pieces of information traversing a given delivery system. Instead, and above all, it had to do with growing awareness of a common, although not always necessarily very tangible, sense of nationhood, and of "one people."[48] It was this exhilarating vision of "the people," the American citizenry deciding in the common interest, which,

44 Blondheim, *News*, 174; Schwarzlose, *Nation's Newsbrokers*, 235.

45 John, *Spreading the News*, 56. As was the case with the postal system, the public space in people's imagination applied mainly to white males, leaving women and blacks outside.

46 Quoted in Leonard White, *The Republican Era: 1869–1901; A Study in Administrative History* (New York: Macmillan, 1958), 14–15.

47 Thompson, *Wiring a Continent*, 252, 445.

48 For instance, putting into words a sentiment that was popular in the middle of the nineteenth century, the lawyer and writer Park Benjamin pronounced that "[t]hough our country covers the broadest area in the world, with the exception of the Russian despotism, capital and labor are bringing together her utmost borders into limits more accessible than were those of England and Scotland a century ago. Besides these grand bonds of the railroads and telegraph and the post, the tides of migration, flowing to and fro, from East to West, from North to South and backward, mingle together our people more and more, and make them one people, one people held together by fraternity and duty and patriotism" (Park Benjamin,

when it started to become more widespread, tended to turn into a demand for homogeneousness.[49] With development of communication networks, the Union would become, it was believed, "more and more one people, thinking more alike, acting more alike and having more one impulse."[50]

Commitment to a social unity, to a truly cultural integration in which divergences and disparities can always be reconciled because of common national ground, was thus fostered and sustained by the ever-extending and deepening infrastructure of centers and links that together formed a system enabling a comprehensive republican public to emerge. Notwithstanding improvements in transportation or communications, however, or perhaps precisely because of those, population was growing more dissimilar and diverse than ever before.[51] Stable structures of meaning emphasizing unity that were linked to agricultural America were becoming transformed into something different through closure of the frontier, the rise of class divisions, rapid urbanization, and a massive influx of immigrants. Extension of markets and enhanced distribution of newspapers increased the possibility for unconnected local conflicts to become articulated as confrontations among political groups on a national level.[52] As conditions of the information economy transformed from an arrangement based on local and regional news distribution to a national information market, patterns that had previously maintained social cohesion became a basis for social diversity and political pluralism.

The horizon of multiplying links formed the prerequisites that to some extent allowed contemporaries to think in terms of cohesion and unity, while at the same time it increased differences in a more complex and uncontrollable social structure. Whereas telegraphic communication indeed improved the means of administering a vast territory with its growing population as a federal unit, perhaps more important it also contributed to undermining the plausibility of thinking about a unified social entity. Therefore, although the telegraph and the press did unify the country, at the same time they also made the differences and

True Patriotism, A Sermon Preached at the Presbyterian Church (July 4, 1851)). Quoted in Merle Curti, *The Roots of American Loyalty* (New York: Columbia University Press, 1946), 116–17).

49 Robert Wiebe, *Self-Rule: A Cultural History of American Democracy* (University of Chicago Press, Chicago, 1995), 124; Meinig, *Shaping of America*, 348. Here, homogeneousness was seen as vital for the sense of nationhood. This view was put into words, for instance, by John Draper, professor of chemistry and physiology in the University of New York, in 1865: "Homogeneousness in a nation makes it conservative, and secures its internal stability by producing a common direction of thought" (John Draper, *Thoughts on the Future Civil Policy of America* (New York: Harper & Brothers, 1865), 107).

50 Editorial in *New York Express* (June 1846). Quoted in Roger Burlingame, *March of the Iron Men: A Social History of Union through Invention* (New York: Charles Scribner's Sons, 1939), 291 and Blondheim, *News*, 190–1. Similar expressions abounded. Thus Asa Child, for instance, praised the network of internal communication that threw "new bonds around the union," and which would "grow stronger as we pursue our forward march" (cited in Russel Nye, *This Almost Chosen People: Essays in the History of American Ideas.* (East Lansing: Michigan State University Press, 1966), 72).

51 See John Garraty, *The New Commonwealth, 1877–1890* (New York: Harper & Row, 1968), 333–4.

52 Christopher Dandeker, *Surveillance, Power and Modernity: Bureaucracy and Discipline from 1700 to the Present Day* (Cambridge: Polity Press, 1990), 59; Patrick Garry, *Liberalism and American Identity* (Kent: Kent State University Press, 1992), 60.

divisions more clearly visible.[53] Naturally, however, this did not prevent new political projects based on the very idea of unity to emerge, as the need to conceptualize social processes under a coherent framework did not go away.

This framework not only implied a shared political system but also constituted the basis on which the political community and its conditions were problematized — that is, presented as an issue. Forming powerfully the communicative condition, the telegraph created, to a certain extent, a context in which community could reflect itself, so to speak, as a distinct political body. It provided the vocabulary and imagery for this reflection, allowing members of society to define experience of the societal and political community in a consistent way that was commonly recognizable. In sum, the telegraph, in particular, constituted not only a tool for creating and fortifying connections but also, and far more important, a system within which these connections could be thought about as a coherent whole.

Universal Language

Thinking about communication in the age of the telegraph was no doubt preoccupied with, even overwhelmed by, the idea of unity. This was partly due to the political concern about national stability as a democracy — a problem to which the telegraph soon came to be seen as providing the solution. During this period, therefore, communication came to be conceived of as a means of engendering a truly nationwide political whole, the functioning of which it would then constantly maintain. Yet, on the other hand, the models and frameworks in which this could take place emerged also from telegraphy itself, as systems and discourses relating to telegraphy provided, with other fields of technologies, forms for conceptualizing the society in a rapidly changing age. What is more, this was not restricted to the Union, for the lines along which the telegraph developed and the logic it followed also seemed to point the way to conceiving of pan-national relationships and the direction in which these would evolve in the future. This was so because transportation and communication appeared to "have lessened the obstacles of time and space which separate the groups of mankind from each other, and have tended to make the whole human race a single unit."[54]

As the extension of telegraph systems continued unabated, it therefore became natural to think that if transportation and communication could harmonize the country by bringing its scattered localities together, why would they not do the same to the rest of the world? The attempt to bring about a harmonious community by means of communication was thus taken to its logical conclusion: the telegraph was seen as being a potentially unifying force among different nations and cultures, a universal language, so to speak, thus evoking the Enlightenment's dream of a continuous extension of the *logos* that would shed the light of reason on dark corners and recesses previously only indistinctly visible, and in this way

53 Marcus and Segal, *Technology in America*, 139; Brown, *Knowledge is Power*, 286, 291–4; Joel Silbey, conclusion in *A Crisis of Republicanism: American Politics in the Civil War Era*, ed. Lloyd Ambrosius (Lincoln: University of Nebraska Press, 1990), 130–1; Blondheim, *News*, 192; John, *Spreading the News*, 13.

54 William Sumner, "Sociology," *Princeton Review* 57 (November 1881): 319.

bring about general harmony among people. Writing in 1858, Charles F. Briggs and Augustus Maverick expressed the belief in this kind of universal brotherhood facilitated by communication:

> It has been the result of the great discoveries of the past century, to effect a revolution in political and social life, by establishing a more intimate connection between nations, with race and race. It has been found that the old system of exclusion and insulation are stagnation and death. National health can only be maintained by the free and unobstructed interchange of each with all. How potent a power, then, is the telegraph destined to become in the civilization of the world! This binds together by a vital cord all the nations of the earth. It is impossible that old prejudices and hostilities should longer exist, while such an instrument has been created for an exchange of thought between all the nations of the earth.[55]

Thus the telegraph appeared, in the language of religious aspiration and secular millenarianism, as a means to building a spiritual community between different nations and cultures, thus ushering in "unity, peace and good will among men."[56]

Hence, the telegraph's spreading out toward foreign countries marked the beginning of universalization of hopes invested in its unifying potential. Behind all this, however, was the expectation of using the expanding telecommunications sphere to advance native political and commercial interests, but this was not considered exclusionary with respect to the pan-national dreams that people entertained. The Atlantic cable project was the foremost intercontinental enterprise at the time, development of which was closely followed and reported on both sides of the ocean.[57] After the line's completion in 1866, which knocked out a project to construct an alternative overland telegraph line to Europe through Siberia,[58] Seward saw it in terms of the expansion of national commerce and political institutions.[59] The physical outward expansion of the telegraph lines, which reflected the rise in importance of commercial and political links to foreign

55 Briggs and Maverick, *Story of the Telegraph*, 21–2. It is noteworthy that the telephone was also described in similar terms later on, and was invested with similar images. Thus for instance, in words of John J. Carty, the vice president of the American Telephone and Telegraph Company (AT&T), "some day we will build up a world telephone system making necessary to all peoples the use of a common language, or common understanding of languages, which will join all the people of the earth into one brotherhood" (cited in Marion Dilts, *The Telephone in a Changing World* (New York: Longmans, Green & Co., 1941), 188).

56 Daniel Davis, *Book of the Telegraph* (2nd edn), (Boston 1858), 44. Cited in Lindley, *Impact of the Telegraph*, 26. Whereas Leo Marx named this language the "rhetoric of the technological sublime," James Carey and John Quirk called it the "rhetoric of the electrical sublime" (Carey, "Technology and Ideology," 206). See also Tom Standage, *The Victorian Internet* (London: Weidenfeld & Nicolson, 1998), 81, 87, 88, 151–2.

57 e.g., John Gordon, *A Thread Across the Ocean: The Heroic Story of the Transatlantic Cable* (New York: Walker & Co., 2002).

58 Reid, *Telegraph in America*, 508–10; Harlow, *Old Wires*, 318–20; Ault, *Wires West*, 76; for a more detailed and recent account, see John Dwyer, *To Wire the World: Perry M. Collins and the North Pacific Telegraph Expedition* (Westport, CT: Praeger, 2001); Rosemary Neering, *Continental Dash: The Russian-American Telegraph* (Ganges: Horsdal & Schubart, 1989).

59 *Telegrapher* 3, (50), (April 15, 1867): 184; also Quoted in Keller, *Affairs of State*, 91.

countries, caused great enthusiasm and optimism among the people.[60] Above all, however, it nourished the conceptions of the civilizating effects of the telegraph.

A contemporary commentator in London extolled the electric telegraph, in 1866, for "it has gathered the civilized world together into instant and direct intercourse; a network of sympathetic intelligence encircles the earth, and we feel as though it had always so existed [. . .]. *Since the introduction of printing, there has not arisen an agency so beneficial and ubiquitous*."[61] People in the nineteenth century, especially in pre–Civil War America, believed that telegraphy would not only bring political coherence to America but also, in parallel with that, improve "the political atmosphere of the nations thus brought into closer moral contact with each other."[62] Capturing this rather naïve universalism of the age, James Reid, a telegraph pioneer who had served as a superintendent in several telegraph companies, held that the telegraph "gave to races of men in various far-separated climes a sense of unity. In a very remarkable degree, not necessarily definable, the telegraph in this mission of union, confederated human sympathies and elevated the conception of human brotherhood. By it the peoples of the world were made to stand closer together."[63]

Also Morse himself believed firmly in the emancipating capacity of the telegraph. "I trust that one of [the telegraph's] effects," he pronounced in 1855, "will be to bind man to his fellowman in such bonds of amity as to put an end to war. I think I can predict this effect as in a not distant future."[64] This was because of the telegraph's capability to make "*one neighborhood* of the whole country."[65] In this way, the formulation originally used to describe the effects of the postal system was rephrased and recontextualized in relation to the telegraph, where it found a new home for a long time. At issue here, as well as in the postal system, was how as vast a nation as America could hold together and function efficiently, avoiding disintegration and chaos. This horizontal perspective framed speech about the telegraph and linked it with the question of maintaining democracy's ability to function. Morse's statement was clearly motivated by this issue of a working order of democracy in a continent-sized nation, but was used to nourish visions of international harmony too. What is noteworthy is that this harmony came to be thought about on the basis of the idea of a neighborhood — in other words, a distinct social entity demarcating what is inside from what is outside — thus giving rise to an image of closure.

In any case, the optimism these enunciations reflected gave expression to general hopes and beliefs of contemporary commentators on the power of the

60 Thus Mr. Latham, for instance, speaking in the US Senate in 1862, expressed the widespread excitement concerning the expansion of the telegraph: "We hold the ball of the earth in our hand, and wind upon it a network of living and thinking wire, till the whole is held together and bound with the same wishes, projects and interests" (cited in Reid, *Telegraph in America*, 508).

61 *Reader* (December 1, 1866); emphasis in original.

62 *Commercial and Financial Chronicle*, No. 1 (July 8, 1865). Quoted in Du Boff, "Nineteenth-Century America."

63 Reid, *Telegraph in America*, 596.

64 "Morse's letter to Norvin Green," July 1855, in Morse, *Letters and Journals*, 345.

65 "Morse's letter to F. O. J. Smith," February 15, 1838, in Morse, *Letters and Journals*, 85 (emphasis in original).

new instrument in advancing peace among men. "The best interests of the world," declared George Wilson, a professor of technology at the University of Edinburgh, "are bound up in [the telegraph's] progress, and its mission is emphatically one of peace. It does not merely speak swiftly but softly, and it offers men a common speech in which all mankind can converse together."[66] The underlying idea of a universal language provided by the telegraph, which would serve as a basis on which different nations could establish a union of peace, was not seen as problematic at the time. This happened because telecommunications did not appear so much as a key military instrument and political tool in the rivalry among the great powers until the end of the nineteenth century, although naturally the telegraph was used for military purposes much earlier. The idea that the language of the telegraph was spoken in a particularly "soft" way became increasingly questionable as the telegraph's position as an adjunct of the process of territorial seizure and occupation and as a strategic instrument in political and military matters in general became clear.[67]

Hopes that accompanied many of the declarations in which the electric telegraph was regarded as "facilitating Human Intercourse and producing Harmony among Men and Nations"[68] ignored the fact that the telegraph was a remarkable social force in its own right. It was not merely a complicated technological device that enabled diverse tasks to be performed much faster and over a much longer distance than before: the telegraph was also a part of a movement that changed the prevailing dynamics in American society. Like every subsequent new form of communications technology, it changed and restructured existing fields of societal relationships and placed them in a new environment. Thus it engendered a renewed form of interaction, a new network for the distribution of information, and a novel matrix for rearranged communication relationships. What is more, it not only connected dispersed communities but also created them, following the railroads in this. The railroads, especially, created the population that used them by enabling communities to develop along its tracks,[69] but this was also

66 Wilson, *Progress of the Telegraph*, 59.

67 Headrick, *Invisible Weapon*, 7; Giovanni Arrighi, *The Long Twentieth Century: Money, Power, and the Origins of Our Times* (London: Verso, 1994), 59. See also Jorma Ahvenainen, *The Far Eastern Telegraphs: The History of Telegraphic Communications between the Far East, Europe and America before the First World War* (Jyväskylä: Gummerus, 1981); Leonard Dudley, *The Word and the Sword: How Techniques of Information and Violence have Shaped our World* (Ann Arbor, MI: Blackwell, 1991). However, as already mentioned, it is clear that the government realized the military and administrative value of the telegraph when it subsidized the construction of the cross-continental line (see Israel, *Machine Shop*, 55; Towers, *Masters of Space*, 97). Although the telegraph was used for the first time for purely military purposes as early as in the Crimean War of 1854–5, it served mainly as an intercommunication and not a tactical device. Its tactical use was not common until the Civil War, which witnessed its first large-scale military application in directing widely separated troops (see William Plum, *The Military Telegraph During the Civil War in the United States, Vol. I* (Chicago: Jansen, McClurg & Co., 1882), 26–8; Russell Weigley, *History of the United States Army* (New York: Macmillan, 1969), 233–4; David Bates, *Lincoln in the Telegraph Office: Recollections of the Military Telegraph Corps during the Civil War* (Lincoln: University of Nebraska Press, 1995 [1907]), 11; Lewis Coe, *The Telegraph: A History of Morse's Invention and Its Predecessors in the United States* (Jefferson: McFarland & Co., 1993), 61).

68 *Daily Chronicle* (November 16, 1847). Quoted in Du Boff, "Nineteenth-Century America."

69 Daniel Boorstin, *The Americans: The National Experience* (London: Sphere Books, 1988 [1965]), 252.

true of the telegraph, which provided the foundation for expanding financial and commercial communities. Finally, it created a whole set of positions, theories, and models that enabled the incorporation of a new field of objects and processes into a more or less consistent body of knowledge that tapped into a common resource of ideas and metaphors. Notwithstanding these matters, the electric telegraph was long understood as a purely enabling system with certain inborn emancipating tendencies, which appeared as an appendage rather than a productive force, an asset rather than a threat.

The reason for this noticeable detachment from power questions lies in a conception of the telegraph as fundamentally neutral in its innermost nature. It was seen as a simple adjunct to existing social practices by improving the way they were carried out, and this was true with regard to not only some individual tasks but also to the functioning of the whole communication economy. This neutrality constituted a precondition for emancipating tendencies: it was the mental terrain that made possible evolution of hopes of a universal spirit of togetherness. When considering contemporary conceptions of the nature of this new form of telecommunication, one can discern at least two distinct yet related features characterizing it:

1) The telegraph constituted a thoroughly instrumental medium for communication, it was thought. Irrespective of its mysterious double nature, which combined the pure miracle of electricity with strange mechanical logic, it was eagerly adopted as a facilitating adjunct to social practices soon after its practicability had been demonstrated. It was an instrumental tool for performing more effectively certain functions that existed prior to its introduction, not a social force in its own right that would engender radically new practices or organizational structures, let alone new ways of being. The telegraph appeared as only an appendage to the immanent societal processes. It was a technology that people could use according to their individual purposes, while its productivity was not yet recognized. The telegraph constituted first and foremost a medium of communication, not a social institution. For, as the issue was put in 1859, "what after all is a telegraph? [. . .] As its name implies, it is a device for effecting the communication of knowledge between distant places."[70] Even 25 years later, it was still considered no more than an "additional and better means for the transmission of intelligence."[71]

Subsequently it turned out that institutionalization of the telegraph resulted in unprecedented accumulation of business power. Growing financial concentration deviated from and simultaneously infringed upon what seemed to be the natural state of affairs, but before this became clear, the operational mechanism of the telegraph was not seen as a productive social force. This was because the telegraph, once combined with the natural dynamics of society, was seen only as improving the effectiveness of society's inborn characteristics. Its exploitation did not aim to change but rather to support prevailing practices, it was thought, without adding

70 Wilson, *Progress of the Telegraph*, 30.
71 Congressional Record, 48th Cong., 1st Sess., vol. 15 (1884): 6041.

any extra features or taking away existing ones. However, if nonetheless it altered something, that was only for society's own good since everything it created was regarded as having already preexisted within it in an inchoate form. Thus, it brought out the inner potentialities of American society, eventually synchronizing its rhythms of life while still maintaining its basic nature intact. It was only through this state of affairs, by being a tool without any positive social nature of its own, that the telegraph came to be viewed as an acceptable political technique for connecting people and regions in a society characterized by the conception of a non-interventionist state. I will return to this theme in the next section.

2) Every communication technology creates its own area of communicativity in the sense that it sets up the particular ideals, definitions, and ways of communication within its domain. In telegraphy an assemblage of these structural determinations was not seen as having any essential influence on the substance of the message itself. In this case, contemporaries distinguished between content and its form of expression, leaving to be discovered the insight that content, like expression, has a form of its own. This allowed them to see the telegraph as an effective way of transmitting preexisting messages without the messages being changed by their transmission. The telegraph was established as a medium for transmitting independent content: what the telegraph transmitted was seen as existing irrespective of its existence.

This was reflected in a tendency to conceive technical improvement in communications not in terms of a new enabling productive principle but instead of face-to-face communication: "The railroad, the telegraph, and the press have virtually brought our citizens [. . .] face to face."[72] The telegraph, as any other communications medium, was seen primarily as allowing communication to take place over distances chiefly in the same terms as those in which it occurred in local, everyday circumstances. Although messages that were sent over the telegraph could not be formulated in as verbose a manner as in interpersonal discussion, and although they had to be transmitted through unfamiliar officials working in the telegraph office, the directness of the messages sent was regarded as remaining unaffected. As the telegraph was almost exclusively a one-way medium, lacking the interactivity present in conversation, the idea that it brought citizens face to face with one another must not be interpreted in terms of conversational criteria, but in some more general context. That context seemed to be the public. The telegraph possessed the capacity to bring individual opinions together in the sense that it enabled modern publicity as a space within which it was possible for conceptions to be viewed alongside one another.[73]

In this sense, the universal hopes invested in the telegraph can be traced back to the project of national unification. As the postal system provided a model through which it could be seen as part of the contemporary communication infrastructure, the telegraph soon became a focal point for a broader and more utopian social

72 Quoted in White, *Republican Era*, 14–15.
73 Blondheim, *News*, 174.

order, because universalization of its potential as a link among individuals, regions, and nations made it a general political technique for community-building. Let me now, lastly, turn to the question of what the cultural self-conception of nineteenth-century America was like, a self-conception that allowed the telegraph to be seen not only as an instrumental and neutral communication technology but also, above all, as a more or less coherent model for social imagination.

Contractual Relations

The nation confronted the question of the telegraph in an era when separation of the spheres of government and individuals, including their mutual associations, defined the political horizon. A strong distinction between state and society was made in the tradition of the classical liberalism of nineteenth-century America. The scope of the state was kept limited by a doctrine of negative state, the central idea of which was that the state was not allowed to interfere with the private sphere. This, for its part, was composed of free, autonomous individuals who were seen as possessing inalienable and absolute natural rights that no government could change. Thomas Jefferson and other political leaders of the early republic held that the purpose of government was precisely to protect these rights, and the theory of natural rights in this way became a foundation for the concept of limited government.[74] The idea of limited government became dominant in the era after the Civil War when the political economy was understood largely in terms of laissez-faire liberalism. The popularity of that concept among academic and popular theorists stemmed mainly from a belief that the social order was governed by an immutable and universal law of nature, that competition was a fundamental characteristic of that law, and, finally, that government was an inefficient agency compared with natural competition.[75]

Free contracts became the organizing principle of private transactions in civil society for most of the nineteenth century, and did not start to change until the industrial age.[76] Liberalism interpreted social relationships as contracts through which individuals sought to achieve personal security and economic gain. Therefore the conception of politics in negative terms, as bearing responsibility for securing civil society's inviolability, must be translated as meaning above all protection of a realm of interindividual contracts. Within this tradition of thinking the contractual relationship was viewed as "rational," as opposed to other, "emotional" relationships such as those relating to family or friendship. The view included a belief in historical progress from an "irrational" society based on custom and sentimental bonds to a rational one in which society would be organized

74 Sidney Fine, *Laissez Faire and the General-Welfare State: A Study of Conflict in American Thought, 1865–1901* (Ann Arbor, MI: University of Michigan Press, 1956), 4; Tariello, *Political Ideology*, 12; Garry, *Liberalism*, 48–50; Howard Gillman, *The Constitution Besieged: The Rise and Demise of Lochner Era Police Powers Jurisprudence* (Durham: Duke University Press, 1993), 24–5.

75 Fine, *Laissez Faire*, 52–5, 58, 131.

76 Morton Horwitz, *The Transformation of American Law, 1780–1860* (Cambridge, MA: Harvard University Press, 1977), 160–210; Kevin Teeven, *A History of the Anglo-American Common Law of Contract* (New York: Greenwood Press, 1990), 179–85, 202. For an early-twentieth-century discussion on the topic, see Roscoe Pound, "Liberty of Contract," *Yale Law Journal* 18 (1909).

in accordance with a model of formal, objective contracts.[77] This conception is epitomized in William Graham Sumner's writing dating back to 1883, according to which in the Middle Ages "society was dependent [. . .] on status, and the tie, or bond, was sentimental. In our modern state [. . .] the social structure is based on contract [. . .]. Contract, however, is rational [. . .] realistic, cold and matter-of-fact."[78] In this way, modern society was viewed as having evolved from a system based on customary status to one based on rational contract.

The doctrine of liberty of contract forged a link that connected the individual liberty of classical liberalism to the market economy. This was true in two senses, for it not only placed individuals with their private associations in the economic domain, but also, most important, became the organizing principle of social relationships, if these were understood as market relationships, as liberalists generally understood them. This link between a political concept of self-government and an economic model of contract was widely thought to have a scientific grounding, as it seemed to demonstrate a natural affinity between legal and economic ideas of liberty.[79] The market mechanism, in turn, was conceived of as forging a tie between private liberty and the general good, as the combined effect of free individual economic decisions was held to advance the common interest.[80] Thus, the liberty of contract constituted a system for mediating on relationships among the right to individual liberty, markets, and the idea of a common good, as all were understood within classical liberalism. In this way, the society in nineteenth-century America was governed by an ethos of freedom of contract: it was, according to legal historian Lawrence Friedman, "the golden age of the law of contract."[81]

It must be borne in mind that the contract institution operated simultaneously on two different registers, for it constituted both a legislative regulation mechanism and a formal model for private relationships. Similarly and concurrently, the telegraph system also operated on two separate levels, in that it not only provided a particular service to be exploited but also constituted a whole framework within which it was possible to view the complex system of communicative relationships as a consistent whole — "a neighborhood." This gives us occasion to investigate the closer connections between the contract and the telegraph, connections that can be viewed at least from the viewpoints of political economy and jurisprudence.

1) Within the liberalist political tradition the telegraph could be conceived as having an inherent relationship to the contract principle. As already mentioned,

77 William Sullivan, *Reconstructing Public Philosophy* (Berkeley: University of California Press, 1982), 15–16, 26, 39.

78 William Sumner, *What Social Classes Owe to Each Other* (Caldwell: Caxton Printers, 1978 [1883]), 23–4.

79 Paul Kahn, *Legitimacy and History: Self-Government in American Constitutional Theory* (New Haven: Yale University Press, 1992), 114.

80 Jeffrey Lustig, *Corporate Liberalism: The Origins of Modern American Political Theory, 1890–1920* (Berkeley: University of California Press, 1982), 2.

81 Cited in Harold Hyman and William Wiecek, *Equal Justice under Law: Constitutional Development, 1835–1875* (New York: Harper & Row, 1982), 42.

the telegraph was conducive to expansion of the market: it rapidly became the foremost medium for commercial transactions whenever speed was a premium value, thus establishing itself as a *de facto* standard in a domain controlled by free contracts.[82] Important, the telegraph was compatible with the prevailing conception of political economy, according to which society was basically composed of independent individuals and their mutual relationships. While the principal social relationship was viewed within political liberalism as being an economic one, the liberty of contract institutionalized a constitutive political mechanism that arranged these relationships into a whole. Similarly, the telegraph provided a standard means for connecting people and facilitating economic transactions rapidly and over long distances. The congruence between the telegraph and the contract principle was commonly recognized, at least among the more business-oriented, as is clear in the following account from 1857:

> Citizens, as a general thing, have no conception of the amount of business daily transacted over the wires. Contracts to buy and sell, pledges of indebtedness, balancing of accounts, all involving millions of dollars, are entered into freely and without fear. From morning till night, day in and day out, are the trembling wires busy with the concerns of an entire nation.[83]

For the telegraph, the world appeared to be composed of different individuals and distinct geographical locations, with the task of this communication apparatus being to forge links among them. So the telegraph's mission was essentially congruent with that of the contract principle: to secure connections within a commercial society that required uniformity and standardization.

2) Moreover, the telegraph was indirectly conducive to the transformation of contract law into an objective legal structure aimed at the regulation of vast, impersonal economic relationships in a modern society. As long-distance trade evolved and fluctuations in supply and prices in different locations were gradually leveled out, due to improved transportation and communication, the amount of impersonal economic relationships formed within civil society increased: the emergence of a national market produced a growing number of transient and anonymous connections among strangers.[84] Thus it became important to create a system to ensure their predictability and reliability, which led to a transformation of contract law from an ethical system to a formal, impersonal jurisprudence. A

82 Alan Stone, *Public Service Liberalism: Telecommunications and Transitions in Public Policy* (Princeton: Princeton University Press, 1991), 25. Unlike in Europe, where it transmitted a much higher volume of government and social messages, the telegraph in America was predominantly a business medium (see Israel, *Machine Shop*, 132).

83 Lefferts, "Electric Telegraph," 251.

84 Christopher Clark, "The Consequences of the Market Revolution in the American North," in *The Market Revolution in America: Social, Political, and Religious Expressions, 1800–1880*, ed. Melvyn Stokes and Stephen Conway (Charlottesville: University Press of Virginia, 1996), 28.

uniform, depersonalized theory of contracts that supplanted the "will" theory — under which every contract was founded upon the mutual agreement, or the "will" of the parties — and thereafter dominated American law, was first articulated in Theophilus Parsons's *Law of Contracts* in 1855. Modern, "objective" contract law, which replaced the face-to-face agreement, was devised and adapted to the particular needs of an industrialized society oriented toward a national market.[85] It was thereby separated from its common-law origin and reconstructed as a modern rationalization of the political doctrine of contractual relationships for the practical needs of an emerging social order. This was linked to changes in cultural structures of meaning and the posing of new political questions, as the telegraph helped to centralize economic power in a way that finally made government interference necessary.

3) Lastly, if the telegraph was conducive to changing the jurisprudence, jurisprudence also changed the legal status of the telegraph. The common-law origin of contract law provided a basic legal structure through which the telegraph itself was regarded as part of the civil sphere, as were any private businesses, although it was practically left free from the common-law obligations applied to common carriers. In the legal framework that gradually evolved to regulate the telegraph industry, the telegraph finally became likened to common carriers and incurred similar responsibilities relating to economic damage caused to customers.[86] Through introduction of a law specific to the telegraph, this technology was for the first time given a juridical identity of its own, and liability for the transmission of telegrams was thus tied to modern jurisprudence in accordance with a general attempt to create a predictable business environment.

Unlike counterparts in Europe, the government of the United States distanced itself from the proprietary ties of this form of communications and let it find its place within the private sphere like any other business.[87] The purpose of politics was not to govern private interests: rather those interests constituted the possibility of politics. The telegraph's influence on changing the distribution of power in society was growing. But that did not pose a political threat because the mechanism of free contract already existed to govern economic relationships among individual interests, while the principle of governmental noninterference was preserved. This was where the central significance of the telegraph for society lay: it was viewed above all as fostering competition by providing a means for forging

85 Horwitz, *American Law*, 184–201; Hyman and Wiecek, *Equal Justice*, 42.

86 e.g., *Telegrapher* 3, (42), (December 15, 1866), 83; Lindley, *Impact of the Telegraph*, 30–70, 125–9, 137.

87 However, closer governmental control of the telegraph was actively lobbied throughout the second half of the nineteenth century, but the bills were not passed in Congress (e.g., Gardiner Hubbard, "The Proposed Changes in the Telegraphic System," *North American Review* 117, (240), (1873): 80–107; Judson, *Government Ownership*; see also Harlow, *Old Wires*, 333–9; Sharlin, *Electrical Age*, 32–3; Fuller, *American Mail*, 172–88; Lindley, *Constitution*; *Impact of the Telegraph*, 108, 125). A good summary is provided in "Selections from Government Ownership of Electrical Means of Communication," see especially the appendixes; Senate Doc. No. 399, 63d Cong., 2d Sess., (Washington, DC: Government Printing Office, 1914).

economic connections and facilitating market transactions. It is not far-fetched, from this perspective, to see the telegraph, with the contract, as an important political technology for organizing relationships among independent individuals and enterprises within the civil society, because the telegraph created a multitude of economic connections over a vast territory without any government intervention, fitting thus inherently into the liberal political tradition.[88]

Yet it was commonly recognized that growing monopolies steadily undermined the assumed natural operation of the economy, understood to be based on self-regulation through free competition and freedom among equal contracting parties. Consolidation of business power in the form of trusts and monopolies was seen as diminishing individual freedom of contract and stifling natural competition. What propelled this process of consolidation was that underlying structural conditions for competition had changed with the growing national market so that competition favored bigger corporations that were able to exploit a scale effect. Or, to put it another way, a company's growth tendency was a reflection of changes that had occurred in the economic system.[89] Although it could be seen as having advanced the common good in pre-war society, competition after the war had plunged industry into serious disorder, which it then had to attempt to manage within itself by seeking more cooperative strategies and mutual interests. One significant consequence of the following large-scale industrial consolidation was the diminution of freedom of contract.[90]

As belief in a harmonious, self-regulating market was gradually diluted by the accumulation of capital it became necessary to rethink the fundamental tenets of the political economy. As emphasis on coordination replaced competition, as was the case in the telegraph industry in the 1860s, and as the complexity and power of industrial systems grew, so the public began actively to demand more control over these powers. The society previously understood as being composed of equal-sized economic units on an equal bargaining level had clearly been replaced by an aggregation of extremely disproportionate economic relationships among different business organizations, as well as between employers and employees.[91] Another phenomenon that encouraged regulation was the fact that

88 See a detailed and insightful discussion on the relationship between the telegraph and liberal governance in Andrew Barry, "Lines of Communication and Spaces of Rule," in *Foucault and Political Reason: Liberalism, Neo-Liberalism and Rationalities of Government*, ed. Andrew Barry, Thomas Osborne, and Nikolas Rose (London: UCL Press, 1996).

89 Of course, there are also interpretations that conflict with this standard view. For instance, William Roy argued in a detailed study that one of the most fundamental factors underlying the rise of the large industrial corporation was a shift in the organization of property, which is ignored in standard efficiency theory (Roy, *Socializing Capital*).

90 Garraty, *New Commonwealth*, 107–22; Peter Hall, *The Organization of American Culture, 1700–1900: Private Institutions, Elites, and the Origins of American Nationality* (New York: New York University Press, 1982), 245; Teeven, *Anglo-American Common Law*, 291.

91 Jefferson and his followers believed in an inherently self-regulating market, consisting of equal sized economic units none of which was bigger or more powerful than the other. Free competition between different interests standing on an equal footing would ensure the ultimate balance and fairness of the market in a predominantly agrarian society. In addition to free competition, wide diffusion of property and a vast open frontier were seen as factors that secured the harmonious working of the market, which is why administration should assume a highly restricted role in economic affairs and particularly should

alongside urbanization slum areas within cities that were densely packed and vulnerable to epidemics began to evolve. Growing pauperism in urban centers on the European model posed an unprecedented problem for American society, one that required a rethinking of the relevance of the concept of minimal government and the policy of nonintervention.[92] Moreover, accentuation of class divisions, growing quantities of immigrants, and the fact that free land available for citizens had run out created more pressure for government intervention. Thus it became gradually obvious that social reform involving expansion of the administrative functions of the state was becoming unavoidable. However, adequate regulation of the telegraph industry was not established on a national basis until 1910, when Congress amended the Interstate Commerce Act of 1887 and declared telegraph companies to be common carriers.[93]

Thus, at the time, many social issues came to be articulated within a dichotomous model in which economic laissez-faire, the tradition of individual freedom and self-determination, and local autonomy were juxtaposed with a need to establish a decent national regulative policy and with pressures to support a more active state. A growing tension emerged between fear of accumulating corporate power and fear of government power strong enough to cope with the rise of an industrial economy.[94] By the end of the nineteenth century, the concept of the positive state had replaced the minimal, noninterventionist understanding of the functions of the state. It also constituted a new view of society. Whereas previously the political community amounted only to the sum of individual interests expressed in society, now it achieved a positive content of its own. Society was now seen as consisting not of atomistic individuals, but of social groups and processes.[95] It was not an aggregation of autonomous individuals, but a positive whole that had an inner

not advance the interests of any single social group in relation to others. Thus, the common-law regime of contract and property, under which social relationships were constructed, was seen as ensuring a fair, self-regulating market environment as long as its "natural" order was not artificially interfered with (Garry, *Liberalism*, 54–7; Gillman, *Constitution Besieged*, 26–7). All these beliefs started to wane by the end of the nineteenth century, hand in hand with the concentration of economic power.

92 Herman Schwendinger and Julia Schwendinger, *The Sociologists of the Chair: A Radical Analysis of the Formative Years of North American Sociology, 1883–1922* (New York: Basic Books, 1974), 38–9; Fine, *Laissez Faire*, 109.

93 Lindley, *Constitution*, 261; *The Impact of the Telegraph*, 30–70, 125–9, 137. Due to his distrust of state intervention and social planning, William Sumner criticized the Act as "the most remarkable piece of paternal legislation that has ever been passed" (cited in Fine, *Laissez Faire*, 90).

94 Garraty, *New Commonwealth*, 110–22; Lindley, *Constitution*, 260–1; Keller, *Affairs of State*, 185, 285, 409.

95 In the early republic, the political community was no more than the sum of the individual interests involved. The whole, the political leaders and commentators of that period believed, was given content and meaning only by its integral parts, and, therefore, it could never be greater than these parts. The "group theorists," in contrast, paved the way for a more collectivist understanding of society by supplanting the idea of atomistic individuals and promoting the notion of groups as the basic unit of society, with the resultant demand to place the public as the object of inquiry, not the individual. Thus the whole was regarded as being more than the sum of its parts. These theorists included thinkers such as Arthur Bentley, Charles Horton Cooley, John Dewey, George Herbert Mead, and Walter Weyl (see Jeffrey Sklansky, *The Soul's Economy: Market Society and Selfhood in American Thought, 1820–1920* (Chapel Hill: University of North Carolina Press, 2002), 131; Tariello, *Political Ideology*, 21, 53, 60–3, 74–81, 103; Richard Hofstadter, *Social Darwinism in American Thought* (New York: George Braziller, 1969), 167–9, 202).

dynamic and inherent laws of its own. This was a moment when society achieved a distinct history of its own separated from the supposed laws of nature. Society not only assumed a more evident character as a political unit but also became a subject for intensive investigation by the new social sciences.

The Nervous System

The telegraph sustained the political self-conception of post–Civil War America that was mainly conceived through laissez-faire liberalism. This is because it created and strengthened private connections within society without governmental interference, thus sharing the same goal as the contract principle — the organizing principle of an economic system based on private transactions. This aspect is crucial, because free, horizontal communication has ever since been associated with telecommunication as one of its key principles. However, what is of even greater significance here is that communication typically necessitates a structural whole, a closure demarcating what is inside and what is outside, which is often understood in metaphorical terms. New communication systems inescapably give rise to conceptualizations and metaphors, in each historical period, that seek to consider and position communication as a conceivable entity and a rational object of thought, and thus try to capture its always elusive "essence" or being. These attempts understandably cannot ever be successful in their goal of presenting what is ultimately unpresentable: if communication can be thought of as having a character of its own, it resides precisely in its nature as denying or repelling any single essence. In any case, these conceptualizations express the necessity to take communication in itself as an object of thought — a critical purpose once the connection between communication and practical governance needs has become evident.

Because they relate to modern communication systems, and because these systems were considered as representing the advanced state of contemporary society, the concepts and images of communication were also inevitably linked to attempts to describe and account for the specific character of the experience of modern society. This is why these vocabularies and imageries must be seen above all as testimonies to social self-conception, for they encapsulate ideas concerning what is the communicative logic of the particular social constellation they belong to. These kinds of representations are needed for making sense of the social whole of communication, and, related to this, for distinguishing what is essential and relevant from what is not. Viewed from this angle, communication becomes closely connected with issues of governance and control, which is of no less importance than the horizontal aspect, and which constitutes the perspective taken in this book.

In this regard, the telegraph assumed a prominent role in the cultural imagination, serving not only as a means but also as a model for social control and thus establishing a critical resource for representations of society in a period that became preoccupied with "science, system and organization."[96] Its logical structure had already been fully articulated by William F. Channing, a remarkable American inventor in telegraphy, in the middle of the nineteenth century. His

96 Keller, *Affairs of State*, 122. See also Marcus and Segal, *Technology in America*, 133.

description of the nature of the telegraph, using a language of organic analogies, was subsequently linked with the Spencerian theory of social evolution that became prevalent in the age of the telegraph. Being in line with Channing's style of formulation, Spencer's vocabulary seemed to provide pertinent conceptual tools for the contemporary conception of telecommunication that was deemed to support splendidly his notion of a biological analogy between an animal and society.

Although an approach to telegraphic communication in late-nineteenth-century America may have been formed on a practical level, the way it was seen as a social form was influenced largely by the widespread biological analogies. The use of the telegraph in national administration and in society at large was not immune to contemporary intellectual discourse: on the contrary, it was mainly understood in terms of Channingian/Spencerian concepts that set the context for the public reception of telecommunication. Discourses on communication and the construction of telecommunication systems should thus not be seen as two separate activities but instead as mutually articulated ones. Spencerian vocabulary was corroborated by the central social role that technology achieved in coordination functions, while the understanding of telecommunication and its potentials were influenced by a climate permeated by this vocabulary of organic systems. Let me first elucidate some central points in emerging social scientific theories relevant to understanding contemporary conception of the telegraph.

Social Organism

Life in early-nineteenth-century America, based on isolated localities with customary rules, could still have been characterized in terms of autonomy, self-rule, and independence. The loose association of isolated localities, which comprised a "segmented society," compartmentalized so that most events began and ended within a local community, had developed hand in hand with the worldview according to which the components of a system of ideas were isolated and abstracted from their contexts, and viewed as independent, homogeneous, and unitary entities.[97] However, development of a national market, and fast acceleration of processes usually characterized in terms of "urbanization," "industrialization," and "modernization," transformed conditions of the comprehension of social reality so that the consequences of human action became intricate, unpredictable, and untraceable to any single source, suggesting a conception of an "organic society." The causes and impacts of individual actions formed chains with distant social processes, became anonymous, and could no longer be handled through traditional conceptual means within a single locality. During this process, social conditions became impossible to describe through the existing liberal principles emphasizing individual and local autonomy. What was needed instead was a new conception of society that would take into account the increasing interdependency

97 See Thomas Haskell, *The Emergence of Professional Social Science: The American Social Science Association and the Nineteenth-Century Crisis of Authority* (Urbana: University of Illinois Press, 1977), 38f; Schwendinger and Schwendinger, *Sociologists of the Chair*, 20; Hofstadter, *Social Darwinism*, 168.

of American society. This changed state of affairs can also be said to have been the reason for the birth of professional social sciences, and, along with them, the whole concept of *the social*.[98]

Clearly, the rise of professional social sciences was possible because of a displacement in traditional cultural structures of meaning — a process dependent on the ever-thickening networks of connections across the country. Social sciences in America began their institutionalization at the same time as the influence of the Darwinian theory of evolution reached them in the 1870s. Herbert Spencer was the most important figure in acceptance of social Darwinism in America. Spencer's works supported a strict interpretation of the politics of laissez-faire. While cautious about analogies between a "body politic" and a "living body," he likened society to an individual organism from a functional point of view because "structures and functions in the human body furnish familiar illustrations of structures and functions in general."[99] As social processes were thus incorporated within an unchanging structure, it became pointless to attempt to modify them in any substantial way. This idea proved useful not only for social scientists but also for conservative politicians, who believed the organic structure could restore social order from the disorganization brought about by the liberal political tradition. Within social sciences, social development became increasingly viewed as following a distinct inner logic that had a precharted course, isolated from historical change.[100]

Spencer expressed his conviction that history was a process in which government control would gradually decline. The distinction he made in his later works between military and industrial types of society was in accordance with this view. Whereas military society was characterized by status and self-preservation, industrial society's primary quality was contract. An individual's life, liberty, and property were no longer subordinated to the state but instead defended by it. As there was no interference by the state in a regime governed by free contractual relationships, survival of the fittest was seen to be ensured.[101] Spencer's huge influence in America helped to establish laissez-faire liberalism as a central view in the political economy after the Civil War. Needless to say, there were also challenging interpretations and counter-currents in this period, but it would still be fair to characterize the laissez-faire interpretation of the liberal political tradition as the prevailing one in late-nineteenth-century America.[102]

98 Haskell, *Professional Social Science*, chapters 1 and 2. See also Dorothy Ross, "The Development of the Social Sciences," in *The Organization of Knowledge in Modern America, 1860–1920*, ed. Alexandra Oleson and John Voss (Baltimore: Johns Hopkins University Press, 1979); Dorothy Ross, *The Origins of American Social Science* (Cambridge: Cambridge University Press, 1991).

99 Herbert Spencer, *Principles of Sociology* (3rd edn), (New York: Appleton-Century-Crofts, 1897), Part 2, 592.

100 Ross, *Origins*, 15, 18. However, when philosophical pragmatism gained ground in the late nineteenth century, the perspective of change that is an essential part of the theory of evolution was taken as one of its guiding principles: society was now perceived as a constantly changing living structure, not as a fixed entity (Hofstadter, *Social Darwinism*, 124–5, 202; Tariello, *Political Ideology*, 72, 84–5).

101 Fine, *Laissez Faire*, 35–6.

102 Fine, *Laissez Faire*, 29–30, 41, 47, 126; Keller, *Affairs of State*, 121.

William Sumner, reputed to be the first American sociologist, mediated Spencerian thinking to a broader intellectual audience and expressed the laissez-faire idea in its most cogent form. He viewed the law of survival of the fittest as constituting a determining context for the development of human societies.[103] It became possible to see the existence of a given social institution or practice as an indication of a successful adaptation to underlying laws of nature, for otherwise it would not have survived. Social Darwinism was in this sense essentially a doctrine that made the prevailing order, viewed as based on an unchangeable course of the universe, legitimate. One could not change these superior, inherent, and inevitable laws of nature that constituted a predetermined structure for human life.[104]

The process of evolution was not only natural but also, above all, an ineluctable one. The progress of human society depended on a selection process based on unrestricted competition. Moreover, because social phenomena were so complex and the human comprehension of them so limited, formulation of unambiguous rules for social reform was ruled out as a matter of principle. It was not possible to change an intricate industrial organization to suit needs evolving at any given time; the best rationale was instead to engineer a more effective adjustment to existing conditions. However, Sumner did not abandon the possibility of sociology, "the science of life in society."[105] He was convinced that "social phenomena are subject to law, and that the natural laws of the social order are in their entire character like the laws of physics."[106] It was the task of sociology to disclose these objective laws underlying the surface events of history; the essence of these, however, could not be changed by human policy.

One of the overriding Spencerian themes that reverberated widely in the writings of contemporary social theorists was the aforementioned conception of society as an organism. Spencer formulated the idea of society as a constantly developing and growing, living organism, uncontrollable by human-made laws. The best analogy for understanding the nature of this being was an animal: "A society as a whole [. . .] represents phenomena of growth, structure and function, analogous to those of growth, structure and function in an animal; and these last are needful keys to the first."[107] By this straightforward biological analogy, Spencer forged a strong correspondence between the structure and function of an animal body and those of a society. Whereas social structure thus became conceived in terms of animal anatomy, social action was understood in terms of animal physiology. This is why

103 "The law of the survival of the fittest," claimed Sumner, "was not made by man and cannot be abrogated by man. We can only, by interfering with it, produce the survival of the unfittest" (Sumner, "Sociology," 311).

104 "Sociology," in William Sumner, *War and Other Essays*, ed. Albert Galloway Keller (New Haven: Yale University Press, 1911), 177.

105 Sumner, "Sociology," 303. See also Hanson, *Democratic Imagination*, 203–5; Fine, *Laissez Faire*, 84; Hofstadter, *Social Darwinism*, 57–62.

106 Sumner, "Sociology," 322.

107 Spencer, *The Study of Sociology*, International Scientific Series (1873), 326. Cited in Schwendinger and Schwendinger, *Sociologists of the Chair*, 40.

it was only natural for Spencer to compare railroad and highway systems with an animal's vascular system, or the telegraph lines with its nerve fibers.[108]

In so doing, he was utilizing the old reservoir of biological metaphors that René Descartes too had found useful when comparing nerves with water pipes, muscles with hydraulic engines, and the respiratory system with a clock.[109] Biological language was also in use when Charles Darwin famously applied the metaphor of organic decay, decline, and extinction in evolutionary accounts that provided a model for describing supposedly similar processes in human society.[110] It was also the same reservoir that German statesman and political economist Albert Schäffle, for instance, had worked on in many of his influential books in the late nineteenth century. Especially in Schäffle's *Bau und Leben des socialen Körpers*,[111] newspapers and other media that transmitted symbolic meanings were depicted as composing a continually growing and diversifying nervous system that would ultimately link every individual.[112] An underlying motive for the use of these analogies by Spencer was to exemplify a conception of society's functional integration, a seamless cooperation by differentiated institutions and operations in highly evolved modern societies, or "superorganisms," as he put it.

The metaphor of an organism provided a representation of society that explained complex societal relationships and processes in terms of a sensible organization: it promoted understanding of the modern society as a consistent whole, suggesting a certain unchangeability and inevitability as the foundation of the political community. Thus, following Spencer, Sumner described sociology as the science that "studies the structure and functions of the organs of human society."[113] Due to its comprehension of society as a body-like system, the science of society, he concluded later, "has worked out a conception of society as a unified whole."[114] The underlying metaphor enabled one to see the society with its functionally specialized institutions and practices, which superficially seemed to be a disordered complex of incompatible and disjointed parts, as an operationally integrated and coherent entity. From underneath the diverse parts and units of society emerged a clear order that rendered the complex and scattered structure an organic being. This being emerged naturally, it was not made. It maintained the same identity throughout its development, and each of its constituent elements

108 Schwendinger and Schwendinger, *Sociologists of the Chair*, 40. See also Jean Quandt, *From the Small Town to the Great Community: The Social Thought of Progressive Intellectuals* (New Brunswick: Rutgers University Press, 1970), 32.

109 John Young, *Cybernetics* (London: Iliffe Books, 1969), 1–2.

110 However, it has been pointed out that Darwin's own theory of natural selection depended on models of social transformation, degeneration, and destruction, thereby completing the organic-metaphor circuit (Whitney Davis, "Decadence and the Organic Metaphor," *Representations* 89 (2005): 131–49).

111 Albert Schäffle, *Bau und Leben des socialen Körpers* (Tübingen: Verlag der H. Lauppschen Buchhandlung, 1881).

112 See Hanno Hardt, *Social Theories of the Press: Early German & American Perspectives* (Beverly Hills: Sage, 1979), 44–57.

113 Sumner, "Sociology," 303–4.

114 William Sumner with Albert G. Keller, *The Science of Society* (New Haven: Yale University Press, 1927). Cited in Daniel Rossides, "William Graham Sumner (1840–1910)," in *The Classical Tradition in Sociology: The American Tradition*. Vol. I, ed. Jeffrey Alexander et al. (London: Sage, 1997), 82.

gained its identity as a part of the larger whole that brought them together.[115] In accordance with this conception, members of the American Social Science Association, founded in 1865, while drawing upon the organicist understanding of sociology, envisioned a natural harmony among people and advocated belief in integration within society as a unified whole.[116] Here the unifying forces of the metaphoric structure became entangled with the language of political economy and the prevailing search for social coherence.

Within this organicist language it is easy to see how the image of a nervous system comprised its very heartland. Analogies of this type — including references to the circulatory system — were frequently used by well-known public figures to describe the postal system for several decades prior to commercialization of the electric telegraph. In a major speech, for example, South Carolina statesman John C. Calhoun termed the postal system the nervous system of the "body politic" in 1817. When an organic analogy was used by Morse in 1838, as he sought congressional funding for the experimental Washington–Baltimore line, he was plainly drawing on these postal analogies.[117] It was William F. Channing, however, who used this analogy in a more systematic way with respect to the telegraph, stating that the electric telegraph "is to constitute the nervous system of organized societies."[118] He was the son of the famous Unitarian minister William Ellery Channing, a prominent thinker in the liberal theology of the time. As already mentioned, it was the elder Channing who, in 1829, stated that the postal system could transform the whole country into "one great neighborhood," thus anticipating the telegraph rhetoric.[119] It has been suggested that William F. Channing's ideas regarding the telegraph were influenced by the discussions taking place in Boston and within his own family concerning a new water system.[120] It is true that the improving infrastructure of Boston — like that of any major city — proved an obvious breeding ground for widespread biological analogies and images. Thus the politician Charles Sumner, for instance, could, in 1847, liken the city's gas pipes to the nerves, its sewers to the veins, and its water pipes to "the arteries through which the life-giving blood is to be poured into our system."[121]

It was the alliance between biological imagery and technological networks that finally embedded the organistic metaphor into the common consciousness. The conception of technological networks — meandering railroad tracks or stretching

115 Kahn, *Legitimacy*, 58–9.

116 Sklansky, *Soul's Economy*, 109–10.

117 "From the enterprising character of our countrymen, shown in the manner in which they carry forward any new project which promises private or public advantage, it is not visionary to suppose that it would not be long ere the whole surface of this country would be channeled for those *nerves* which are to diffuse, with the speed of thought, a knowledge of all that is occurring throughout the land, making, in fact, *one neighborhood* of the whole country". ("Samuel F. B. Morse to Francis O. J. Smith, Feb. 15, 1838," in Morse, *Letters and Journals*, 85; John, *Spreading the News*, 10).

118 William Channing, "On the Municipal Electric Telegraph: Especially in Its Application to Fire Alarms," *American Journal of Science* 13 (1852): 58.

119 John, *Spreading the News*, 13.

120 Joel Tarr, Thomas Finholt, and David Goodman, "The City and The Telegraph: Urban Telecommunications in the Pre-Telephone Era," *Journal of Urban History* 14, (1), (1987): 38–80, 76n57.

121 Cited in Tarr, Finholt, and Goodman, "The City," 76n57.

telegraph lines — as the vascular or nervous system of modern American society presented a startling vision to contemporaries. Transportation routes made the functioning of a societal organism possible, for traffic over rails and roads presumably performed the same kind of task as circulation of blood in an animal body. On the other hand, telegraph lines seemed to enable a conscious coordination of society in a similar way that the nervous system transmitted the impulses of the higher nerve centers of an animal to its organs. In this way, the telegraph provided a supposed central agency with a means to direct the functioning of society, and, at the same time, made visible a principle of conscious command.[122] This vision was in wide use among contemporary popularizers of science and other semi-academics, in addition to scholars within the institutionalizing social sciences, who were interested in current technological developments.[123] So it was not meant as an overstatement when C. R. Henderson, an associate professor of sociology at the University of Chicago, claimed in accordance with contemporary understanding in 1897 that "telegraphs and telephones [. . .] are really the nerve system of the social body, the material means of welding this nation and all nations into one spiritual community."[124] The telegraph helped society as a whole to better fulfill its true nature by equipping it with an effective control mechanism. It let society, it seemed, control its own being.

However, it is important to note that the development of transportation and communication, on the one hand, and neurophysiology, on the other, coincided in time with mutually constitutive effects: scientists studying organic and technological communications systems inspired one another. In this way, the metaphors used worked in both directions, because while technological instruments and new communication devices influenced the way scientists understood nervous activity and nerve-impulse transmission, physiological studies, in turn, inspired communications engineers.[125] This effective relationship is also manifested in the

122 Secular millenarianism and organicist analogies often coalesced in this manner when new communication methods were devised. This can clearly be seen in the following quotation by Mr. Crittenden, speaking on the proposed "oceanic telegraph" between the coast of Newfoundland and the coast of Ireland at Congress in 1857: "I should be glad to see [the telegraph] pervade the whole world. I think it would be of advantage to mankind if these lines of communication could be established everywhere, so as to reach civilized man in every position in which he is placed on the globe, so that they should pervade the whole system just as our nerves and sinews pervade the human body. I think it would have a great tendency to the advance of mankind, and contribute greatly to a general sensibility all over the world, that we might know all that was doing in every part of the earth — the right and the wrong, to feel for each other, to make mankind move all together and in harmony with the events of the age in which they live" (Congressional Globe, 34th Cong., 3d Sess., (1857): 423).

123 See Charles Rosenberg, *No Other Gods: On Science and American Social Thought* (Baltimore: Johns Hopkins University, 1976), 4–6.

124 C. R. Henderson, *The Social Spirit in America* (New York: Chautauqua-Century Press, 1897), 96–7.

125 Laura Otis, *Networking: Communicating with Bodies and Machines in the Nineteenth Century* (Ann Arbor, MI: University of Michigan Press, 2001), 3–4, 11–13. Linda Simon has added that rather than being parallel or metaphorical, language used to describe nervous action and the telegraph reflected common perception that animal electricity and artificial electricity were identical in nature; Linda Simon, *Dark Light: Electricity and Anxiety from the Telegraph to the X-Ray* (Orlando: Harcourt, 2004), 305n25. This view was encapsulated, for instance, in *North American Review* 87 (October 1858), which Simon cites: "By the connection now established between distant cities and opposite hemispheres, we have but arrested, for a special subdivision of one among its many departments of service, a force which throbs from zone to zone,

ideas developed by Channing, as he was also a trained physician, with a fascination for the application of electricity to medicine. He published a book on the subject in 1849, writing of the principle of "vitality." This principle worked on both the nervous system and the living tissues, modifying "existing electro-chemical affinities as by the power of electricity itself."[126] By combining his medical training with his interest in electricity he was the first to systematize the organic vocabulary with regard to the telegraph. According to the *Telegrapher*'s account, the mechanism of the human system suggested to Channing the separate "signal" and "alarm" circuits that characterized his plan. This made "the analogy with the functions of the motor and sensitive nerves of the animal organization [. . .] complete. The central office is the *brain*, the wires the *nervous* system."[127] This comparison was elaborated in a more explicit form in a paper he delivered to the Smithsonian Institution in 1855. "The electric telegraph," Channing remarked,

> is [. . .] the nervous system of this nation and of modern society, by no figure of speech, by no distant analogy. Its wires spread like nerves over the surface of the land, interlinking distant parts, and making possible a perpetually higher co-operation among men, and higher social forms than have hitherto existed. By means of its life-like functions the social body becomes a living whole, and each of its new applications marks a step in the organization of human life.[128]

Here the pre-Spencerian use of the biological analogy did not serve as a retrospective rationalization of the telegraph system but instead as a conceptual prototype for its technical structure. It did not *explain* but actually *directed* thinking about and the construction of the electromagnetic telegraph, and of the sphere of communication it enabled.

The anatomical comparison largely laid a foundation for contemporary interpretations of human-made technical systems, as it seemed to provide a good way to conceptualize their social function. This is also why it made sense to attempt to extract the inner laws of nature and translate them into some practical, applicable rules, for this was seen as the only way any human innovation would work. An invention that conflicted with these laws would have no chance to survive. This is exactly where the ingenuity of Channing's recipe lay: in order to be able to simulate nature's logic of operation, one has first to systematize its central principles. This is also the reason he thought himself entitled to deny that his interpretation was a pure analogy. Instead, through it, so he seemed to think, nature revealed its innermost essence.

leaps from sky to earth, darts from earth to ocean, courses in the sap of the growing tree, runs along the nervous tissue of the living man, and can be commanded for the speaking wires simply because it is and works everywhere."

126 William Channing, *Notes on the Medical Applications of Electricity* (Boston: Daniel Davis Jr., 1849), 3–5, 11.

127 *Telegrapher* 1, (2), (October 31, 1864): 14.

128 William Channing, "The American Fire-Alarm Telegraph," *Ninth Annual Report of the Smithsonian Institution* (1855): 148.

Although it appeared the right tool to decipher nature and the mechanisms of some major technological systems, the analogy remained somewhat isolated until it was subsequently incorporated into a more general theory in Spencer's interpretation. The way Channing used it in relation to his telegraph was broadened and connected to a coherent theory of social development. With the spread and popularization of the Spencerian theory in the United States, the biological analogy was entrenched as a general truism in public awareness.

The transformation of American society from a "segmented" society with isolated and self-sufficient communities to an "organic" society, in which the regions and localities were tied together by a process of growing interdependence, constituted a precondition for a change in the conception of telecommunication from a purely neutral form of technology to the nervous system of society.[129] This marked a new stage of cultural sentiment through which communication in the age of the telegraph assumed a clear, distinctive, and positive character. This is not to say, however, that a total break in conception of communication occurred either. As the effects of social transformation became visible, the telegraph inevitably got a more positive identity, but, at the same time, it largely retained its instrumental nature in the sense that it was held to be directable by human decisions: although becoming more clearly a social power in its own right, telecommunication was seen as remaining under the control of the people whose practical affairs it helped. The changes the telegraph contributed should still not be underestimated. It became entwined with the historical vicissitudes of the nation in that it was conducive to processes that changed the conditions for thinking about not only different technological systems but also the whole social organization in terms of a network — or better, a nervous system.

In new social circumstances in which anonymous chains of actions penetrated distant regions, the telegraph came to be regarded, alongside the railroads, as the most visible representative of this novel logic of operation. The telegraph's interlinking power formed a sensible model for thinking about increasing interdependence and interpenetration of distant regions and actions. Not only was it conducive to facilitation of nationwide circulation of actions and implications but also it came to constitute a model on the basis of which this phenomenon could be conceived. Thus, the telegraph had two simultaneous effects: it was at the same time a historical force for societal transformation and a model or metaphor for conceptualizing the new societal logic. The nervous system as an analogy for the telegraph provided an excellent conceptualization for the latter purpose. In particular, it seemed suited to describing, first, the telegraph's use and position in tasks of the national and municipal administration and, second, the thickening networks of urban communication, which represented points of concentration of social power.

The analogy of a nervous system was not only an attempt to describe the

129 It was Émile Durkheim who described societal transformation as a movement from "segmented" to "organic" society. Haskell, *Professional Social Science*, 38n13, has briefly discussed this conceptualization in relation to the idea of interdependence.

prevailing structure and function of the social organization in terms of communication but also, and above all, an attempt to think about the "whole" of communication as it was understood at the time. By means of the analogy it became possible to conceive of the "essence" of communication — or the realm of what could be called communicativity — which presupposed thinking about its conditions. Through the idea of a nervous system, the social organization appeared as having a distinct organicist foundation based on laws that science could unravel. Yet, within a still broader context, the biological analogy facilitated perception of the political community as a fundamentally communicative project, the task being to forge connections that would foster the feeling of unity. In sum, then, the biological analogy constituted a relationship through and in which thinking about communication was given a definite shape and form. These sorts of forms, after their formative period, directed and structured thinking about society by providing stable, unifying frameworks. With respect to thinking about communication, the idea of the nervous system proved a lasting conceptual whole that not only constituted a structure for communication but also, more important, formed the perspective within which communication could be viewed as an object for thought in itself.

Electric Nerve Centers

Although it was potentially a general communication system, the telegraph, partly due to its expensive use, was largely related to governance and coordination. The logic or nature of the diverse functions it added to different operational domains varied remarkably. Despite these disparities, an underlying conceptual image of the telegraph as the nervous system of society constituted a unifying conception for communication, as already discussed. National and city governments together with the metropolitan press and news agencies and telegraph offices can be regarded as realizations of the idea of the nervous systems of a nation, in that they were the primary intersection points of inflows and outflows of telegraphically transmitted information. As municipal telegraph systems had established a basis for a centralist scheme, this idea — a dominating center and local offices arranged in a hierarchical manner with constant information traffic among these locations — provided the context within which both to think about and to effect the practical organization of governance in terms of communication on an increasingly regular basis. This is why we have to study local systems more closely, because they, especially the police and fire emergency systems, came to be referenced in thinking about more general societal processes too.

The telegraph issue needed to be addressed not just as a general concern as to its position in the liberal political discourse. More practical and specific questions also arose as to its role in governance within an organizational framework. The question of possible uses of the telegraph for government purposes, however, was not a particularly urgent one. Although administration was weak and lacked appropriate means for enforcing legislation, the telegraph did not immediately appear as a mechanism to meet these needs. This was partly because of a common fear of strong central government and partly because the telegraph was first seen rather as a kind of scientific tool with no inherent practicability for serious applications. Efforts to connect this system to government objectives increased steadily, as its role in a national communication infrastructure gradually became clear. Eventually, the diverse administrative tasks of the government came to be closely connected to the telegraph.

This communication system was connected mainly to functions that served a number of communicational and scientific purposes within administrative institutions. It was placed in the service of politicians, for example, for keeping in touch with their constituents; of diplomats, for correcting misapprehensions; of scientific and technical government agencies, such as the United States Coast Survey "in charting weather reports and spreading warnings of approaching storms, calculating longitude, establishing correct time at various points throughout the country, and in registering astronomical observations."[130] In this way, scientific practices, as well as political ones, helped to reduce uncertainties and increase predictability.[131] The telegraph became an important part of gradual standardization of different spheres of life, coalescing with other technical developments to make the social more predictable, stable, and governable.

The telegraph developed into a common instrument in administrative departments in federal and state governments, being used for ordinary departmental business that required dispatch. By 1860, the General Land Office was sending telegrams routinely to correspondents five days late in responding.[132] After the telegraph had been established in government use, arrangements for a comprehensive telegraph system for government service in Washington were being advanced, making its position as the leading political center of the national body gradually more visible and unambiguous.[133] One of the central advantages of the telegraph in governance was in enhancing the ability to coordinate diverse operations and geographically scattered units under centralized management and in controlling the whole intricate system more effectively. It was largely due to this capacity that advocating the popular ownership of the telegraph appeared to be a legitimate option at the time in that it seemed natural to embrace it within one centralized management structure.[134]

The telegraph accelerated communications between central government and its representatives across the country, and between headquarters and smaller units within a region, thus strengthening the importance of a central executive agency. As administrative historian Leonard White pointed out, relationships between headquarters and the field no longer depended wholly upon a lengthy correspondence consuming weeks or months, since now field information, thanks to the telegraph, could be transmitted straight away to Washington and responses

130 Thompson, *Wiring a Continent*, 252.

131 Prescott, *Electric Telegraph*, 252.

132 Senate Ex. Doc. 1, 36th Cong., 2nd Sess., 76. Quoted in Leonard White, *The Jacksonians: A Study in Administrative History, 1829–1861* (New York: Macmillan, 1954), 487.

133 Congressional Record, 48th Cong., 1st Sess., vol. 15 (1884).

134 Du Boff, "The Telegraph," 267; Judson, *Government Ownership*, 88–91, 104–15, 166–7.

dispatched on the same day.[135] So close did the connection between the telegraph and central administration become that, according to the Circuit Court in the Pensacola case in 1875, "the functions of the Government either in war or in peace could not now be carried on without [the telegraph's] use."[136]

The new dimension of rapidity that the telegraph introduced not only strengthened but also reorganized administrative organization. Institutionalization of the telegraph in governmental organizations was part of the growth of professionalism in administration that involved the concept of administration as properly businesslike rather than political in nature. This was connected to the emergence of formal bureaucratic control: local means of governance were substituted by bureaucratic state organizations that were separated from party politics and the market economy.[137] Now government was equipped with better means for monitoring the varying dynamics of the society, although control on the whole, as well as improvement of the government's administrative effectiveness, remained minimal by modern standards. This was mainly because the political climate did not allow government to take full advantage of available technology and strengthen its control capabilities. Any attempt to increase the power of central government was confronted by a resistance stemming from the political tradition that relied on the principle of self-governance. However, it was precisely the idea of a dominating center that was gaining strength as the telegraph became common in governance practices.

Municipal authorities, in particular, were predominantly at the leading edge of telegraphic systems development. This, in fact, was the field of administration in which the concept of the nervous system proved most apt, providing a model to be seriously considered also in national administration: management of space and movement in a city came to be discussed as a model for governance of the country. One question that required urgent attention was control of space in rapidly growing urban areas, as many American cities were beset by problems caused by a fast-increasing and densely packed population. Such cities became explosion points for urban riots, fires, and epidemics around the middle of the nineteenth century. What has been mentioned with regard to the changing administrative structures applied particularly well to municipal organizations: implementation of municipal telegraph systems, introduced as technical adjuncts to city governments in a battle against urban disorder, was part of a professionalization of fire departments and police organizations. Partly thanks to the telegraph, both groups were reorganized into more unified and efficient bureaucratic organizations. The telegraph made coordination of units faster and more effective, permitting continuous control of city space to become gradually institutionalized in

135 White, *Jacksonians*, 487.

136 The Pensacola Telegraph Company v. The Western Union Telegraph Company, 1875. Quoted in Lindley, *Constitution*, 229.

137 White, *Republican Era*, 388; James Beniger, *The Control Revolution: Technological and Economic Origins of the Information Society* (Cambridge, MA: Harvard University Press, 1986), 280; Dandeker, *Surveillance*, 60, 100–1.

administrative machinery. In this way, the telegraph equipped government with a means that enabled development of an organized practice to penetrate society on a more detailed level than had been possible before.

William Channing seems to have been the first to introduce a model of a telegraph system tailored to municipal governance needs. He did that in 1845 in an article in the *Boston Daily Advertiser* in which he described the general principles of his system for applying the electromagnetic telegraph to fire alerts, a system that subsequently became the basis for the American fire-alarm and police telegraph systems.[138] In 1851, he submitted a detailed plan with Moses G. Farmer for a telegraphic fire alarm to the city government of Boston; the plan was adopted and put into operation in 1852.[139] Under this scheme, Boston was first subdivided administratively into seven fire districts. Then a central station at the city hall was connected to 19 belfries and 39 alarm boxes throughout the city, thus establishing a center with incoming and outgoing lines. Turning a crank inside the box transmitted a signal to the central station identifying the box and the fire district in question. On this basis, the central office operator could signal identification of the district to the bells as well as that of the originating box to all alarm boxes so that by listening to the nearest box the firemen could locate the fire.[140]

What is of interest to me here is that this proposal did not remain an isolated plan: it became a blueprint that has subsequently been relentlessly reproduced and imitated, installing the signal box as a fixture in American cities. With this simple and elegant design Channing introduced what later became a prototype for many similar systems in different cities, forming what was called the "Boston system."[141] It consisted of three parts: an "Electric Centre, where the batteries and certain instruments, with a single operator, are placed," and "two classes of circuits," the one being "Signal," "by which the intelligence of a fire is communicated to the Centre," and the other being "Alarm," "by which the impulse is sent out to the machinery by which the alarm-bells are struck."[142] The benefits of this system, according to Channing's account, included "instantaneous," "universal," and "definite" communication.[143] Philadelphia and many other large cities adopted the Boston system by 1860.[144]

What emphasized the status of this system as a distinct diagram of communication, unlike existing models, was its capability to formulate an original blueprint for the needs of urban governance. For Channing, the conception of telegraphic communication had to be rethought since existing systems could not serve as models for a city telegraph. Tasks and aims related to the former could

138 William Channing, "Morse's Telegraph for Fire Alarms," *Boston Daily Advertiser and Patriot*, (June 3, 1845).

139 Prescott, *Electric Telegraph*, 237.

140 Tarr, Finholt, and Goodman, "The City," 56–7.

141 Israel, *Machine Shop*, 111; Tarr, Finholt, and Goodman, "The City."

142 Channing, "Municipal Electric Telegraph," 60.

143 William Channing, "Report Concerning a Uniform System of Fire Alarms," *City Document No. 42, City of Boston* (March 24, 1851), 9; Channing, "Municipal Electric Telegraph," 64.

144 New York City used the telegraph in some form for fire alarms already in 1849, but shifted to a variant of the Boston system in 1870 (see Tarr, Finholt, and Goodman, "The City," 77n66).

not possibly form any substantive part of the latter, which had a distinct logic of its own and should therefore be categorically differentiated from the "common" telegraph. To clarify this claim, he stated in a lecture given at Yale in 1852 that the telegraph had two functions: "The first is to interlink distant communities, and to this its applications have been hitherto chiefly confined. The second which is hardly less important is to bring into coöperation the members or parts of single communities."[145] Keeping precisely this second function as his target, he planned his telegraph system as a concentration of points that were close to one another rather than a dispersed distribution of distant points, thus replacing a network based on a number of centers by one that had only one.

Instead of the "linear" connections that characterized the long-distance telegraph, which set out to conquer unbounded expanses, municipal systems would operate in loops, circles, and intersections: unlike long-distance links, municipal links did not embody a single direction or relationship. Whereas long-distance telegraphy was based on straight lines generating reciprocal relationships across wide areas, an intraurban system operated on the basis of contacts among different hierarchical strata within a multiplicity of possible points of interaction. The space for intraurban systems no longer consisted of a heterogeneous diversity, but rather of a hierarchical density. Thus, whereas the common telegraph is a "line" of telegraph,

> [t]he municipal telegraph is the application of the telegraph to a *surface*, making it cover a space with telegraphic nerves and papillæ as thickly as required, to furnish a complete organization. The common telegraph connects *distant* points, [. . .] the more distant the better to illustrate its character. The municipal telegraph contemplates the linking together of a multitude of *near* points, the nearer the better to illustrate the peculiarity of the system. The common telegraph connects two independent centres of life and activity. The municipal telegraph connects a multitude of subordinate points with one centre, and makes the position of those points dependent upon the centre and the needs of the system.[146]

Contrary to Channing's conception, however, the difference between the two systems was perhaps not so absolute after all. In fact, distinct telegraph lines were transformed into a network form as early as in 1846, thus facilitating the centralization process that came to characterize the industry.[147] Furthermore, the city telegraph only concentrated all the connections and relationships of a long-distance system into a single, geographically limited unit. With the help of a system of preselected codes, urban telegraph could squeeze all transmissions into a multilevel complex. Thus, what was initially formed through straight lines was now organized in folds and circles; namely, on a principle of concentricity. This, however, brings out two fundamental yet interconnected characteristics that make

145 Channing, "Municipal Electric Telegraph," 58.
146 Channing, "American Fire-Alarm Telegraph," 149.
147 Blondheim, *News*, 55–6, 190.

the city telegraph stand apart from the common telegraph: the idea of a dominating center and the alliance between communication and control.

1) To be sure, the common telegraph already facilitated centralization in industries such as the railroads, in which it became crucial to control the far-flung systems from a central location. In fact, the systemic nature of the telegraph in itself stimulated the growth of centralized control.[148] Yet it was the city telegraph in which this principle found its foremost expression within the new domain of telecommunication. This is because the more a single intraurban system came to be defined by circles and loops, the more the idea of a strong center became reinforced. After all, it was the task of the municipal telegraph, more than of the common telegraph, to connect a multitude of subordinate points to one center that governed the whole system. The multiple points of connection served as sources through which information from any part of a city could be sent to a center in which it was analyzed and from which an alarm would then go out. "The agent at [the central] office would have it in his power to communicate directions to all the stations, and every alarm of fire might be made to pass through the central office before being communicated to the stations generally," as Channing explained in 1845, adding that there was "no end to the useful modifications of which this design is susceptible."[149] Diverging from the contemporary political order favoring decentralized government, Channing thus came to articulate a powerful center-based view, built on his telegraph design.

2) However, Channing's system also had a second important characteristic that distinguished it from the prevailing conceptions concerning the common telegraph: while the latter was primarily about communication, his system was predominantly control-oriented. His key insight was to complement the vision of the telegraph as the nervous system of the living nation by emphasizing the role of what he called "the motor nerves" of the animal system. For him, the nervous system was not only about communication — what he likened to the sensitive, "afferent" nerves. The point was to see the system as a functioning whole, and to include the "efferent" nerves in order to complete it. In the common telegraph, "communicating intelligence between distant places performs the function of the *sensitive* nerves of the human body. In the fire telegraph it is made to act for the first time in its *motor* function, or to produce effects of power at a distance."[150] This task depended on the brain, or what corresponded to the central station in his design. For this reason this second characteristic is inherently interlinked with the first one, as it is due precisely to the control function that "there should be a Centre."[151]

148 Israel, *Machine Shop*, 24.
149 Channing, "Morse's Telegraph."
150 Channing, "American Fire-Alarm Telegraph," 153.
151 Channing, "Municipal Electric Telegraph," 60.

A center had to be presupposed since the area to be governed constituted a system, and every system required, it would seem, a central organ. Moreover, as the telegraph system corresponded naturally to the nervous system, it had to represent a control mechanism for the social body. The underlying biological analogy directed thinking about the system, for just as coordination of a living body depended on the activity of the brain, coordination of a city system was thought to require a commanding center. Through this center, the whole could not only function consistently but also become aware of itself.

> [The municipal telegraph's] function is not to connect distant towns or independent centres of life and activity with each other, but it is to organize a single city or town so as to bring every subordinate part into relation with its centre of government and direction. Its purpose is to multiply points of communication, to cover the surface of the municipal body as thickly, if you please, with telegraphic signalizing points as the surface of the human body is covered with nervous extremities or papillæ, the whole being intelligently connected into a system by which the municipal body shall understand itself in every part, and shall have a common life and vital functions for its own essential purposes.[152]

In this quotation, the idea of a self-conscious system was given an explicit articulation. The whole is first divided up by the telegraph into different segments, after which every single part of the system is locatable. When all the distinct sections have possibly been subdivided further into different districts, all connected to local centers, the segments can finally be linked to the main center of the system. In this scheme, the center would, in principle, be capable of governing every part of the whole, according to hierarchically arranged relationships. Moreover, the connection between the governed parts and the center enabled the whole to be aware of its own being, as it were. Now also the idea of a self-governing whole could be expressed in terms of communication. The center can be cognizant of events going on in different parts of the system, while individual points of contact distributed throughout the whole provide the means for the whole to attain a certain level of technical self-reflexivity. This is why Richard Vaux, a mayor of Philadelphia, could say that "the telegraph is not only the voice, but the eyes of the executive department."[153] All component parts could thus be geared to the interests of a commanding center, which, for its part, could be aware of the whole because of the density of connections between it and the various parts of the social body.

The telegraph also figured centrally in a wider range of urban governance tasks than was covered by the fire-alarm telegraph. Police telegraph systems constituted the second critical application of municipal telegraphy that was often based on the Boston system. In these systems the telegraph was used in controlling urban

152 Channing, "American Fire-Alarm Telegraph," 149.

153 William Philips and Charles Robinson, *Philips' & Robinson's Municipal Telegraph*. No information about the publisher (1857), 10–11.

disorders such as riots and crime, thus specifically contributing to the governance of the city space of the largest metropolises, which were beset with unrest and frequent riots from the 1830s through to the 1850s.[154] In addition to enabling the transmission of information on criminals between different cities, it allowed better coordination within the police force by linking the stations to central headquarters within one city. The New York City Police Department installed a telegraph system between headquarters and ward offices in 1853.[155] In Boston, a police telegraph system was designed using Channing's fire-alarm system as a model.

Although the dissemination was not as rapid as with the fire-alarm telegraph, the police telegraph soon became an indispensable instrument in police organization. By creating direct connections between police districts and headquarters, it lessened a need for frequent meetings between captains of districts and the police commissioner. It enabled coordination of action among separate stations and was adopted subsequently for connecting patrolling policemen to their stations by using call boxes that were connected either to the precinct house or to headquarters. By 1864, New York City was divided up into 58 signal stations:

> From the central office [. . .] wires are extended to every part of the city, connecting with the signal boxes and alarm bells, and returning again to the central office. [. . .] Each signal box is furnished with a telegraph key for police purposes, and, by a simple set of signals, any policeman can communicate with the central office from any part of the city.[156]

Channing's idea of organizing the system according to "signal" (*afferent, sensitive*) and "alarm" (*efferent, motor*) circuits was the underlying model here too, although in contrast to the original fire-alarm system the signal boxes contained a dial enabling policemen to pass a specific request to headquarters.

It allowed centralized control of the city space and of police forces, thus representing with its fire-alarm systems the proverbial central organ of the social body. The *Scientific American* resorted again to the old analogy when describing the combined telegraph and telephone system in 1881:

> When the entire area of the city shall have been covered by the system the analogy between the civic organization and the nervous organization of an individual animal will be curiously complete. The civic organization will become sensitive, so to speak,

154 Tarr, Finholt, and Goodman, "The City," 61.

155 *National Intelligencer* (March 1, 1853). Quoted in White, *Jacksonians*, 487.

156 *Telegrapher*, October 31, 1864, 14. By 1890, however, criticism started to mount concerning the "antiquated" state of the police telegraph system in New York ("An Antiquated System," *New York Times*, (August 10, 1890)). At that time, the city of Brooklyn was divided into 19 police precincts and the station houses were all connected by telephone with police headquarters. "By means of a switchboard at the central office any two or more station houses can be placed in communication, and the officers in command can talk to one another. In sending out general alarms, giving descriptions of fugitive criminals, missing persons, lost children, &c, all the nineteen station houses can be connected with the central office, and the message can be sent to all the precincts simultaneously and without repetition. It takes only from three to ten minutes to send a general alarm message from the Central Office to all the precincts" (ibid.).

> at every point, and the transmission of intelligence therefrom to the brain and subordinate nervous ganglia — that is, the central and district police stations — will be practically instantaneous.[157]

Principles of transparency from the center and of hierarchical relationships in relation to information transmission came to be understood not only as characterizing technical features of new systems but also, above all, the structure of the police organization itself.

According to the *Telegrapher*, "through [the police telegraph] the Superintendent transmits all his orders to subordinates, and he can, if he chooses, sit quietly in his arm-chair in the Mulberry street police palace and direct the whole immense machinery under his charge." The superintendent, however, was not the only user of the system: "The Police Commissioners and their chief clerks also use the police telegraph in the transmission of orders, and the transaction of nearly all other official business."[158] With this system, the police could improve control of city space qualitatively, as telegraph wires twisted a network around a city, allowing subtler and more exhaustive control over its interstices. Like the fire-alarm system, the police telegraph divided a city up into sections, created relationships among stationary points of contact and station within a section, and finally connected all sections into a unified structure. "By this means," declared Judge Conrad, ex-mayor of Philadelphia, in 1856, "I am enabled to command the entire [police] force at any time. Movements of the police can be directed at every minute in the twenty-four hours by an unseen order." "Now, the entire police force has but one soul," he summed up, "and that soul is the Telegraph."[159]

In addition to its application against social disorders, the New York police telegraph system was used for sending "all notifications to the coroners" and those "in regard to offal, dead animals, nuisances, etc.," for "restoring lost children, recovering lost or strayed horses, arresting deserters from vessels, protecting shopkeepers and others against newly-issued counterfeits and other descriptions of bad money," and for "the suppression of riots, the dispersion of mobs and the allaying of popular disturbances of all kinds."[160] Thus, the superintendent of the police telegraph could announce in 1869 that with the aid of the telegraph, "128 lost children have been restored to their parents; 170 lost, strayed, or stolen animals have been returned to their owners; 58 vehicles have been similarly disposed of; 542 dead animals have been reported to the sanitary police, and 15 alarms of fire have been sent over the wires."[161] Police telegraph systems were adopted at a somewhat slower pace than the fire-alarm telegraph, and the number of systems

157 "The Chicago Police Telephone and Patrol System," *Scientific American*, (April 23, 1881): 240, 258. Quoted in Tarr, Finholt, and Goodman, "The City," 64.

158 *Telegrapher* 3, (54), (June 15, 1867): 228.

159 "Letter from Richard Vaux, Mayor of Philadelphia." In Philips and Robinson, *Municipal Telegraph*, 10, 9.

160 *Telegrapher* 3, (54), (June 15, 1867): 228.

161 House Ex. Doc. 1, Part 3, 41st Cong., 2d Sess., (1869): 1123.

started to accelerate only after 1882.[162] Although the telephone rapidly replaced the telegraph as a central police communication medium, the police telegraph constituted an effective aid to the police, being, according to a contemporary commentator, "equal to a model guard of two hundred and fifty men."[163] It formed an essential component of a technical basis for coordination not only of the police force but also of governmental institutions responsible for the internal safety of society, of control over and on behalf of the subject population.

As mentioned, creation of an improved technical basis for coordination and control was part of professionalization of the police and other administrative institutions, which, in turn, was linked to a process in which local governance practices came to be rationalized and supplanted by national bureaucratic control of formal state organizations. The construction of a new centralized police organization was a reflection of this. Intermittent and reactive control, concentrated mainly on individual incidents or areas, was replaced gradually by preventive, detailed, and regular supervision of society.[164] This was connected to the application of scientific techniques to the practice of governance, to establishment of separate functionally specialized public institutions, and to general improvement in government data-gathering, storage, and retrieval. Given this, the US Census Bureau started to collect large masses of comparable information. The expansion of the practice by governmental bureaus of collecting large amounts of comparable statistical information was spurred by the growth in public finance and the demands of taxation and the military.[165] The Bureau of Statistics, for instance, was organized as a branch of the Department of the Treasury in 1866, and the first Bureau of Labor Statistics was created in Massachusetts in 1869.[166] With the help of new bureaucracies, government started systematically to gather information on its subject population and of societal processes after the Civil War.

These practices developed hand in hand with the increasing use of statistics among social scientists. The conception of statistical work changed in the Gilded Age from a descriptive statistics of social groups and a rudimentary science of natural laws to a scientific method proper for analyzing diverse social phenomena following the German model. In this way, the notion of social processes following unchangeable laws of nature was gradually replaced by an idea that those processes had an inherent dynamic of their own, brought about by collective human activity. A positivist conception of natural law was still widely maintained in social sciences, though.[167] However, it now became possible to articulate societal development in its own terms — separated from any metaphysical laws of nature — composed of a complex set of human actions. Thus, society achieved a history

162 Tarr, Finholt, and Goodman, "The City," 63.

163 "Letter from Judge Conrad, Ex-Mayor of Philadelphia." In Philips and Robinson, *Municipal Telegraph*, 9.

164 Dandeker, *Surveillance*, 110–23.

165 Raymond Gettell, *History of American Political Thought* (New York: Century, 1928), 615.

166 Keller, *Affairs of State*, 102, 168.

167 Ross, *Origins*, 59–60.

of its own, gradually undermining the traditional ahistorical framework through which its development has previously been conceptualized.

Systematized Assemblages

Whereas introduction of the telegraph can be said to have had a profound influence on national and municipal administration, the effects on the institutionalization of communication, on its structure and limits within the emerging communicative formation, were not less sweeping. As in the former case, the biological analogy provided a cogent framework for thinking about communication. In fact, this analogy, which was embodied in the very model articulated by Channing, while becoming generalized and extended in scope retained the same form of a nervous system. At the same time, communication, understood as a means of control and system-building, came to be more closely connected with the systematization of processes, activities, and social structures — a tendency with which late-nineteenth-century Americans became preoccupied.[168] This was related not only to the appeal of centralization and the search for presumably natural hierarchies but also to the growing proclivity to conceive of the market society as based on interdependence and reciprocal roles rather than individual ownership and autonomy, as was traditionally understood in the context of political economy.[169] The increasing systemicity created preconditions for the substitution of the classical conception of society as a neutral domain based on competition among free, isolated individuals by the notion that it is based on cooperation and reciprocal relationships in an ever more complex whole. Thus the systematization of communication not only reflected but also directed the way communication came to feature in social thought and action.

The telegraph had a strong systematizing tendency in communication that had an undermining effect on the decentralized information distribution that the government had supported. As centralized communication systems were established as hubs for the circulation of information they were also translated into models for the formalization of communication within a logical structure. The conditions that paved the way for the generalization of Channing's scheme, in broad terms, prevailed on at least three dimensions related to the development of concentrated communication flows. First, the forming of big corporations and monopolies along with the emergence of nationwide markets; second, the development of numerous intraurban telegraph-based communication and information services to help commercial, financial, and domestic sectors; and lastly, the metropolitan press and news agencies, accompanied by an emerging nationwide system of news wire services, growing into national information hubs. The remarks made about these processes are naturally not intended to imply that centralization, as it appeared across societal sectors, was a consequence of these developments alone, or that it somehow applied only to the communicative sphere and related

168 Marcus and Segal, *Technology in America*, 133–40.
169 Sklansky, *Soul's Economy*, 111.

technologies. All I am claiming is that they helped to transform the communication model by strengthening centralizing forces and tendencies.[170]

1) The telegraph reinforced forces for economic concentration and monopoly organization while at the same time improving the functioning of markets, as these two developments were mutually complementary processes in the rise of managerial capitalism. As the telegraph favored users who needed rapid long-distance communication, companies were prompted to exploit it to its full advantage by expanding their scope of activity and operating on a national basis.[171] During the late nineteenth century, large business organizations came to dominate the economy as a result of improved transport and communication. As already mentioned, the telegraph came to support a concentration of economic power under centralized direction, contrasting with its previous role as an agent in helping the progress of an ideal fostered by the Jacksonians, in which small enterprises competed against one another within formally equal conditions.[172]

2) The second dimension in which the telegraph's centralizing power was manifested implied the rise of intraurban telegraph systems, a new industry that began in 1867, which became a principal area for the development of telegraphy after the Civil War. After a variety of systems based on pre-assigned functions and automatic signaling mechanisms had been developed, specialized uses of the telegraph multiplied and became broader in scope. Urban metropolises became concentration points for a number of different telegraph-based information services, such as the gold indicator and the stock ticker, which enhanced and speeded their internal communication.[173] They constituted rapidly growing arenas on the basis of which ever-denser communication networks were interrelated with evolving social power formations that extended their influence all over the country.[174]

3) Lastly, and perhaps most important regarding communication, the metropolitan press and news agencies grew, alongside national governmental bureaucracies, as principal information intersections, the true concentration points of the transforming communicative assemblage.[175] They showed the direction in which communication would find its future structure, and the way in which it would have to be arranged on the basis of centralized organization hand in hand with

170 Thus William Sumner, for instance, maintained that the new communication technologies were behind the centralization of social control, because communication provided the capabilities of "close, direct, and intimate action and reaction between the central control and the distributing agents" (cited in Hardt, *Social Theories*, 217).

171 Du Boff, "Nineteenth-Century America," 576–7; Du Boff, "The Telegraph," 253–4, Lindley, *Impact of the Telegraph*, 7, 133.

172 e.g., James Hurst, *Law and Markets in United States History: Different Modes of Bargaining among Interests* (Madison: University of Wisconsin Press, 1982), 20.

173 Reid, *Telegraph in America*, 602–34; Tarr, Finholt, and Goodman, "The City," 46–8.

174 e.g., Allan Pred, *The Spatial Dynamics of U.S. Urban-Industrial Growth, 1800–1914: Interpretive and Theoretical Essays* (Cambridge: MIT Press, 1966).

175 e.g., Blondheim, *News*; Mott, *News*; Pred, *Spatial Dynamics*.

the general consolidation of bureaucratic control. A strong linkage between the newspaper press and development of the telegraph increasingly rearranged the whole existing system for collecting, processing, and distributing news, which was based on editorial exchange of news. Now the telegraph became a necessity for individual newspapers, with an increasing volume of information that had, prior to the telegraph, been available only intermittently, now being systematically transmitted to newspapers.[176] Let me elaborate a little further on this issue.

The telegraph system, providing rapid one-way communication over distance, favored centralized news-gathering agencies. The New York Associated Press, one of the best-known national syndicates gathering and distributing information to the press between 1848 and 1897, was exactly this kind of center: it received pieces of news from all over the country, which, after a process of selection and editing, were sent back in standard form to local newspapers. Thus, it was based on a similar scheme to that originally formulated by Channing for a fire-alarm telegraph, in which information came in from numerous points in a system and was gathered at its center, where it was analyzed and finally sent back to the public in a prearranged fashion. In the Associated Press system, the content of information transmitted — namely, diverse instructions, reports, and alarms — was simply replaced by individual pieces of news reports. However, whereas in the fire-alarm telegraph a single, hierarchically organized system was the basis for operations, the newsbrokerage system extended its operations across the nation and incorporated a large number of local centers and regions. In this way, the principle that Channing had successfully applied to cities by translating and folding the linear lines of long-distance systems into circular and hierarchical ones was reversed and stretched again to cover the entire area of the social body.

Thus, the newsbrokerage system was structured according to a more extensive scheme. It was based on interlinking a number of distinct points or poles, thereby creating a network for constant circulation of information. It operated on a principle reminiscent of Channing's system, but expanded it again to a multipole organization. Although the poles in the system were interactive in the sense that they both sent and received information, only one center dominated the configuration and determined the principles on which it worked. Thus this centralized system of newsbrokerage displayed a twofold structure. In the first place, the numerous points of contact served as sources of new information to be edited at the center, which needed active contribution from its local agents in the form of continuous information feed. In the second place, the peripheral points of distant localities were transformed from active sources to passive recipients of information, becoming terminuses for final news sent from the center. So while the first part of this structure was based on an arrangement in which a large number of individual sources fed information into a single center, the second one turned this structure around as it embodied the unprecedented principle of the mass media:

176 Blondheim, *News*, 56. A contemporary account of the news wire system is provided, for instance, in W. F. S. Shanks: "How We Get Our News," *Harper's Magazine* 34, May 1867, 511–22).

the narrowing of the range of news sources, and the corresponding widening of the range of passive reception.[177] To put it in yet another way, while the first, inflowing movement provided new, heterogeneous, and potentially incommensurable information for the system, the second, outflowing movement provided edited, congruent, and essentially identical information.

As in fire-alarm systems that concentrated control of an urban area at a single point, the Associated Press system centralized news-gathering and delivery into a single structure. As speed improved and technology developed, however, it became possible to let a particular piece of news be posted directly to the whole network from a remote location, thus increasing the system's flexibility with important messages. Hence, the headquarters' role changed from that of a collection and delivery center to that of a supervisor or regulator of the system, retaining control over policy but dispensing with the practice of centrally editing the incoming flow of raw news.[178] The newsbroker service nevertheless remained a centrally organized and professionally controlled business that reached an ever-increasing reading audience over ever-lengthening distances, with the center assuming an unquestionable position of command with respect to the rest of the system.

Given this, what can be said about a change that occurred in the communicative assemblage? The order between regional powers lost its stability as communication and transportation systems coupled those powers with a close interrelationship. Hence, the meaning structure that was based on a predictable network of forces had to be recast for a new environment in which previous relationships had lost their validity. What was at play here was not only an expansion of private economic power but also a transformation of the underlying conditions of social life. As has been shown, in the course of this process the idea of a centralized hierarchy was strengthened, hand in hand with the growth of professionalism in municipal schemes of governance and the creation of large-scale bureaucratic structures in public administration. Having unleashed — together with improvements in transportation — accumulation processes for capital and power, and enabled construction of centralized systems of coordination and communication, the telegraph provided an illustrative and exhilarating model for thinking not only about the needs of governance and control but also increasingly about the underlying rationale that characterized the rapid transformation of contemporary society. Thus introduction of the concept of a center in American political culture was linked — at least from a communicative perspective — with this twofold operation of the telegraph as a technology and a conceptual model.

Of course, however, the rise of the telegraph institution represented only one albeit significant historical process related to the idea of a center becoming formulated. Like any tool, the telegraph was a metaphor of itself, symbolizing the

177 Harold Innis, *The Press: A Neglected Factor in the Economic History of the Twentieth Century* (London: Oxford University Press, 1949), 36–7.

178 Richard Schwarzlose, *The Nation's Newsbrokers: Vol. 2, The Rush to Institution, from 1865 to 1920* (Evanston: Northwestern University Press, 1990), 117, 146.

activities it enabled.[179] However, its character as a model was not limited to the way it was used as a technology but was instead more profoundly entwined with cultural self-conception articulated in an explicit form in the Spencerian theory of society. While supporting free individual activity and minimal government, it also came to conceptualize control based on the role of a central organ of the social body. It was introduction of the telegraph as the nervous system of a modern society that embodied and made visible this principle of control; and this, for its part, paved the way for similar arrangements and expressions in social thought. In this way, although not the only cultural model for it, the telegraph enabled conceptualization of a more justified view of a hierarchical, centralized system that was needed in social control.

Certainly, this ran counter to the liberal tradition of American political thought. Yet it instilled a vision of the functional advantages of a hierarchically organized whole in thinking about social questions — a vision that gained more ground in the progressive era. It provided a model to cope with interlinking clusters of problems like those that ensued from growing immigration and the worsening condition of cities. This was because of a conception that what a municipal telegraph system did to a city, a social system organized on similar principles could do to the whole country; namely, divide it up into distinct units and classes, set up boundaries between different areas and districts, form horizontal and vertical relationships between them and the center, and finally render this hierarchical network governable, at least in its essential aspects, from a single location. On the other hand, like the news wire service, it could organize channels, hubs, and intersections for communication flows, establish institutions for systematic gathering, restoring, and retrieving differentiated types of information, and, finally, connect all the flows into a dominating center.

To be sure, this way of thinking characterized more the aims enacted in Europe, especially those of Prussia and France, than those of America. For the latter, acentric systems rather than centralized ones constituted a foundation for political structure. One should not, however, presuppose too sharp a conceptual distinction between these organizations, in line with the famous division between *Gemeinschaft* ("community") and *Gesellschaft* ("society") proposed by the sociologist Ferdinand Tönnies, for they were inherently related to each other. Acentric systems were not independent of centralized ones, which, for their part, could not contain the necessary movements of disintegration within them. Thus, for example, the news wire service enabled and was conducive to the processes that produced a number of heterogeneous information systems and increased the horizontal flow of communication. These systems inevitably influenced the way the news wire service worked by, at the very least, constituting a thick texture of institutions and practices that served not only as a background but also often as a source of counter-truths for national news. However, in many cases it was still

179 Joseph Weizenbaum, *Computer Power and Human Reason: From Judgement to Calculation* (San Francisco: W. H. Freeman, 1976), 18.

against the very figure of centralized control that these heterogeneous practices came to be measured and thought about.

Communicative Whole

Even if society did not, after all, turn into an immense nervous system, facilitated by the metal tentacles of the telegraph, the very idea doubtless forged a new kind of thinking on social questions and on the broader level of the political community. Combined together in a symbiotic relationship, models provided by some networked systems, worsening social problems, and theoretical discourses of the Spencerian variety seemed to justify demand for a stronger centralized political government. Ever more complex chains of problems and the accelerated pace caused by technological development fostered a growing inclination for similar kinds of arrangements to those that had successfully been applied in private businesses to maximize the scale effect in nationwide operations; namely, governance of a variety of processes within a more homogeneous, powerful, and unified administrative structure according to the model of the railroad and telegraph industries.[180]

Underlying this demand was a perceived need for the rational treatment of problems and the "scientific" coordination of the economy and of the nation.[181] This, in turn, prepared a terrain for technocratic thinking to take root later in the following century. It articulated a model on the basis of which political questions could be handled successfully, it seemed, as technical problems. The apparent efficiency of organizational arrangements prepared the ground for an emerging approach that sought to formulate a technical model for management of political processes. It received its most famous theoretical embodiments in Taylorism and the technocracy movement, the influences of which, notwithstanding their limited periods of active practice, were not transient but left a lasting trace in common ways of conceptualizing political questions as well as in the history of communication seen as an instrument of control. It is here that the role played by the telegraph is significant, as the model it lent to the service of powers championing a centralized decision-making structure was also conducive to technocratic phrasing of non-technical problems together with results produced by institutional reorganizations.

Of highest interest for me are not the possible effects of the telegraph on specific schemes for thinking about political problems, although this is certainly not unimportant. Rather, my interest lies in communication's position as a precondition of the political community and in the telegraph's central role in the new communicative sphere. No doubt without the telegraph, communication would

180 My intention is not to claim that these industries provided the only model of political centralization, or even that the incentive for this originated exclusively in the corporate world: the changes in the telegraph industry, as well as in other business sectors, reflected the general bureaucratization of economic and political power in American society (see Israel, *Machine Shop*, 151). Both political and business discourses were mutually articulated within a broader social transformation toward more unified and centralized organizations, each providing models and metaphors for the other.

181 Israel, *Machine Shop*, 55–6, 183, 187.

not have become politicized nor conceptualized to the extent it was at the time. It was the telegraph that produced an unprecedented positivity in communication; an awareness of the powerful role of communication was only intensified during the era of the telephone. By the end of the nineteenth century, people started to pay attention to a reciprocal relationship between political questions and communication practices that had an effect on each other, and thus to the productive influence of communication systems to society. Now communication could be regarded as an explicit and positive tool to be used for political purposes — an observation that incorporated it into the tradition of political thought.

What is significant in the history of communication from the perspective of this book is that "communication," in the age of telecommunication, depended on a larger conceptual whole. Thinking about communication was inherently connected to the idea of a prior whole that enabled it as a social experience. This was realized at least in three related aspects.

1) The idea of unity that marked political discourse provided an environment in which communication was not only politicized but also conceptualized in terms of its own positivity. The theme of national unification served as a perspective within which many disparate communicative events and processes could be conceived of as being subservient to a single objective or purpose, a principle that would give communication a consistent structure.

2) Moreover, as a result of the telegraph and a subsequent new economic identity for communication, attention was focused on its mechanisms, movements, and forms. As thinking about communication was motivated more by practical than theoretical interest, not least because of the inchoate state of scientific communication discourses, commercial organizations that produced and directed these flows became the reference points for "communication" as a social phenomenon. Communication strengthened economic consolidation and the rise of industrial capitalism, but it was this process, more than any other, that created the social interest in it and brought it out as a distinct conceptual entity.

3) Lastly, concurrently with the formation of new administrative bureaucracies and the attendant increase in professionalism in governmental organizations, communication came to be functionalized in terms of novel streamlined governance systems. It was mainly due to the telegraph, which facilitated this process, that the alliance between communication and control became settled. Communication has been associated with social control ever since, although naturally it cannot be reduced simply to this function.

The whole that seemed to both enable and set the limits for thinking about communication in these processes can be named. It was the nervous system, which constituted largely the framework for the discourse of "communication," understood as a means for and basis of certain overarching political, commercial, and administrative projects. Thus, insofar as communication was viewed as a socially effective object in itself, it is possible to argue that this took place mainly in terms of the biological analogy, which conceptualized communication as a social mechanism and rendered it the driving force for various forms of social self-representation. The analogy of the nervous system, specifically, laid a foundation for the idea that communication could be subordinated to a single law or principle. In the course of this idea's development, communication became regarded through the themes of ordering and planning as an object to be organized according to consistent functional principles.

The biological analogy turned this fact into the basis for political order, and provided a model and legitimation for political attempts to seize "communication" in the service of national unification and social control. Under the sphere of influence of this powerful image of communication as the vital system of modern society, communication was not only rendered a "whole" but also made a rational whole since the lines and routes of communication were seen as constituting ultimately an orderly system serving specific social functions. The biological analogy sought to rationalize networks of exchange and organize the tracks and connections of information flows under the same conceptual whole. This is a theme, however, that found its greatest expression during the telephone's heyday.

3

The Speaking Machine

The ideas concerning communicativity underwent a change at the beginning of the twentieth century. This was mainly because of the coincidence and interweaving of two distinct fields of power. The emergence of new conceptions and representations was occasioned, first, by the rise of rational planning, which in turn was an outcome of the coming together of efficiency doctrines and scientific method. All this culminated in the principles of scientific management, which, for its part, paved the way for the triumph of a hierarchical system as a paradigmatic organizational technique. Second, the technology of telephony opened up a whole new era for communication and constituted the chief framework for these processes of rationalization and systematization seen from the communicative viewpoint. Together they came to draw up a novel diagram of communication.

The position of the telephone[1] is central in the modern history of communication. Forming the first real interpersonal telecommunication system, it assumed at the beginning similar functions to the telegraph, but soon turned, after the creation of networks predicated on telephone exchanges, into a general conversational medium.[2] So if "[e]lectrical communication was the function of the telegraph," then "[e]lectrical intercommunication of anything like a general character was reserved for the telephone."[3] As it did not require expertise or, eventually, intermediaries to operate, as was the case with the telegraph, the telephone made possible instant interpersonal communication, and not, as the telegraph did, only interregional communication. "The fact that any normal human being could telephone," said telephone historian Arthur Pound in 1926, "dedicated the

1 The telephone was described by Alexander Graham Bell, in his letter "to the Capitalist of the Electric Telephone Company" in 1878, as "an electrical contrivance for reproducing in distant places the tones and articulations of a speaker's voice, so that conversation can be carried on by word of mouth, between persons in different rooms, in different streets, or in different towns" (cited in John Kingsbury, *The Telephone and Telephone Exchanges: Their Invention and Development* (Arno Press, (1972 [1915]), 89). For an enthusiastic account of the experiments conducted by Alexander Bell and Thomas Watson at the office of the Boston Rubber Shoe Company, see "The Telephone," *New York Times*, (February 3, 1877). See also "Sent by Telephone," *Boston Daily*, (February 13, 1877), and "Voice Telegraphy," *Boston Daily*, (February 24, 1877), for reports about the lecture in which Bell explained his invention. Bell also gave a second lecture three months later, which was also eagerly reported, for instance, in "The Speaking Telephone," *New York Times*, (May 19, 1877).

2 Colin Cherry, "The Telephone System: Creator of Mobility and Social Change" in *The Social Impact of the Telephone*, ed. Ithiel de Sola Pool (Cambridge: MIT Press, 1977), 114–16.

3 Kingsbury, *Telephone*, 77.

telephone to the masses beyond all recall."[4] This personal accessibility brought about the immense constitutive effect the telephone had on the functioning of society. This provided the prerequisites for it to be considered not only as an instrument enabling new forms of organization to be founded but also as a central institution by way of which the social organism could organize its inherent being according to a conscious rationale.

This is why the telephone system, "a great communication machine,"[5] was not just an institution among others. Its significance resided in the fact that it allowed — so it was thought — social institutions to adapt themselves to the changed circumstances of modern society: by rationalizing operations and increasing their efficiency. On the other hand, it was to a great extent the very improvement in communication and transportation that set in motion the dissolution of the previous social order, a process we understand as modernization. Thus the telephone system cannot be considered as only having made a "contribution [to] the functioning of [social] institutions,"[6] as many commentators commonly tend to think in line with functional theories, because this system was a precondition for existence of these institutions in their modern state. But even this is not all. The telephone system can be said not only to have constituted a precondition for collective institutions and their possible interrelationships that together made the society what it is, at least from the communicative perspective, but also, more accurately, to be the expression and mark of the temporal unfolding of an industrial order insofar as the latter is based on communicative relations. This is why communication systems do not contribute to, reflect, or follow the social as if being external to it; rather, they are the very loci of the social, and thus expressions of the unfolding of what the social "is" — an unfolding that consists precisely of this process of expression.

The telephone, after its use was socially established, created conditions for interpersonal communication that transgressed and restructured communicative practices being formed in large organizations. Unlike the telegraph, in which

4 Arthur Pound, *The Telephone Idea: Fifty Years After* (New York: Greenberg, 1926), 10. See also Ruth Cowan, *A Social History of American Technology* (New York: Oxford University Press, 1997), 161. The advantages of the telephone over the telegraph for local business were, according to two notable telephone developers, as follows: "1. that no skilled operator is required, but direct communication may be had by speech without the intervention of a third person; 2. that the communication is much more rapid, the average number of words transmitted a minute by Morse Sounder being from fifteen to twenty, by Telephone from one to two hundred; 3. that no expense is required either for its operation maintainance or repair. It needs no battery, and has no complicated machinery. It is unsurpassed for economy and simplicity" (Gardiner G. Hubbard and Thos. A. Watson, May 1877, cited in Kingsbury, *Telephone*, 67). The telephone was seen to have just the same advantage over the telegraph that motor transport had over the railroad: "You start at your own end and go straight through to your destination, whereas the telegraph compels you, like the railway, to journey to a fixed point in order to set a new piece of machinery in motion" ("The March of the Telephone," *Manchester Guardian*, (October 5, 1923)). Yet the other side of the story is that, compared with the transmission of telegraph signals, the transmission of telephone signals required huge amounts of power and theoretical understanding, which explains the long time separating the primitive urban services in 1877 and the first Atlantic cable in 1956 (see Peter Hugill, *Global Communications since 1844: Geopolitics and Technology* (Baltimore: Johns Hopkins University Press, 1999), 54).

5 Herbert Casson, *The History of the Telephone* (Chicago: A. C. McClurg & Co., 1910), 156.

6 Cherry, "Telephone System," 125.

the ticker, on the one hand, and the news service, on the other, had created a medium to be used for broadcast purposes too, the telephone was mainly an interpersonal medium. While allowing people to communicate with one another without going "through channels,"[7] the social impact of the telephone system has at the same time been conducive to the formation of extensive hierarchical systems. Although there is much truth in the argument that the telephone was not important to the establishment of internal lines of communication as a managerial tool, being mainly a mechanism to extend informal communication,[8] it formed at any rate a conceptual model for this purpose. The telephone, besides enabling free, spontaneous communication that was — and still is — commonly associated with the qualities of a democratic society,[9] also constituted a context within which thinking about communication was tied to the values of regularity, functionality, and rationality in the early twentieth century.[10] Although horizontal communication never ceased to unfold, systemic, hierarchical communication rose from the sea of horizontal communication and was widely regarded as paving the way for the future of communication. It created not only areas of organization but also an image of the community itself, by providing an idea of how it functioned as a whole.

Thus, although it had a remarkable role as a precondition for horizontal communication, in the end the telephone had what was perhaps an even more important function as a foundational model for administrative power and social control. As its position as a condition of these practices became clearer, so the promotion of horizontal relations and the distribution of information, as outlined by the progressives, started to yield to the demand to found a law, understood here as an objective structure presumably beyond political contention. This law, to determine and regulate relations and practices within the purview of a given system, was already a prominent part of the formation of the new socioeconomic order in terms of systematically applied methods, hierarchically arranged organizations, and routinized practices. However, it was not until the telephone was established as the main form of electric communication that law, as opposed to education and the distribution of information, became an undeniable basis of

7 Joshua Meyrowitz, *No Sense of Place: The Impact of Electronic Media on Social Behavior* (New York: Oxford University Press, 1985), 322.

8 See JoAnne Yates, *Control through Communication: The Rise of System in American Management* (Baltimore: Johns Hopkins University Press, 1989), 21.

9 e.g., Page, "Social Aspects" (in *Modern Communication*), 6, 18; Pool, *Forecasting*, 86. This close connection between communication and democracy has already been noticed by Alexis de Tocqueville and Émile Durkheim (see Dario Melossi, *The State of Social Control: A Sociological Study of Concepts of State and Social Control in the Making of Democracy* (Worcester: Polity Press, 1990), 4, 56, 102). On the subject of the telephone, the historian Marion Dilts wrote in 1941 that "inherent in our [American] character is a firm belief that true progress depends upon a free exchange of knowledge and ideas, and our telephone system makes it possible for us to put this faith into practice" (Dilts, *Changing World*, 37).

10 In her history of electric communication, Carolyn Marvin mentions briefly this tension between democracy and control that is inherent in modern communication technologies (see Carolyn Marvin, *When Old Technologies Were New: Thinking about Electric Communication in the Late Nineteenth Century* (New York: Oxford University Press, 1988), 100).

the social order.[11] While law is understood here as a set of predetermined rules and methods, communication as a precondition of big bureaucratic organizations can be similarly conceived of as a relation in which function and parties are predetermined with respect to the system through which the relation is realized. Communication here is, above all, subordinate to the system that employs it. In other words, it receives its meaning from the functions of the system, and whose status and purpose have thus been formally pre-given, which is not the case with horizontal relations.

In what follows, I will analyze the dimension wherein communication was connected to law, understood as a formal mechanism to control interrelations within a given system, instead of to free horizontal relations. The role of communication in social control, once established, necessitated the search for laws to govern it, presumably constituting universal and indisputable logical structures to be uncovered through scientific research. But the existence of these structures is not something that can be disentangled from the historical unfolding of communication itself, as if external to it. On the contrary, communication as a social phenomenon is nothing other than the historical process through which a community constantly reorganizes itself as a community.

Communication lent itself as a seemingly neutral, accurately definable mechanism to the service of powers seeking rational governance and determined social relations. Here the application of scientific research to social questions cannot be separated from rational reform and subsequent projects, from Taylorism to the technocracy movement and beyond, for social rationalization. Communication took a form in which it became both an object of scientific concern and a useful tool in the prevailing institutional practices grouped on and around the foundation of the aim of rational order. Of course, communication can be understood as a law only in the same social practices and the field they form, wherein the social sphere, too, becomes thought of, in particular, as a resource that is capable of being rationalized. Thus, the role of communication in the modern world[12] — in its position as a certain kind of law — is indistinguishable from the history of modernity.

In this chapter I will deal with the systematization of communication in relation to both the systems of and the discourses on the telephone. Along with the development of telephony, the theoretical cornerstones of this process, from the perspective of this book, included the rise of the "scientific method" in social thought concurrently with the establishment of professional social sciences and communications engineering, the emergence of the technocratic branch of the progressive movement, and the formulation of Taylor's scientific management, which systematized efficiency doctrines. Let me now turn to the theoretical

11 See Barry Karl, *Executive Reorganization and Reform in the New Deal: The Genesis of Administrative Management, 1900–1939* (Cambridge, MA: Harvard University Press, 1963), 9.

12 By *modern* — in the context of American society — I have referred in this book to the industrializing social order as breaking with the agrarian system based largely on relatively self-sufficient localities with local governance. As regards the communicative sphere, this change can be seen as concurrent with the introduction of telecommunications technology. The concept also refers to a particular way of conduct, defined by an obsession with rationalization, systematization, and control.

tradition that formed the conceptual terrain for thinking about communication, after which I will discuss, in the next part of this chapter, the forms and modes of communication as realized in certain individual systems, although these should not be thought of as isolated but, on the contrary, as interdependent, inherently connected phenomena.

Taylored Systems

It is possible to see "the speaking machine," as the telephone was often called in the contemporary press,[13] as having formed a speaking machine of the whole society at the end of the nineteenth century. This is because it provided a pattern not only for communication but also for "social system" viewed as a whole.[14] The telephone can be seen as an intersection of powers in and through which both modern communication and industrial order — which are not to be treated as separate from each other — assumed their shapes. It is my intention here to bring out the path of development through which communication, seen as both the object and means of political projects during the telephone's heyday, was given pivotal position in the representations of society.

In the course of this development the structural conditions of communication — that is, the ontological dimension of communicativity — has been understood in terms of a whole characterized by growing systemicity. In this heterogeneous but unidirectional project, wherein communication was given ever more precisely formulated meanings, the connection through which thinking about communication does not deal with its determinable characteristics but rather with what in communication precedes all particular communicative meanings and its articulable properties begins to be perceived within an ever more technological context. This is why the line of development, brought up here, from Lester Ward through Frederick Taylor to the technocracy movement and further to cybernetics, is important. For it was in these discursive frameworks where the foundation for considering communication primarily as a question of technique, method, or law was laid — a way of thinking that marked the modernity. It was in and through these forms that communication found its modern political appearance.

Communication as a social question, during this period, was posed by way of these instrumental settings mainly through two distinct, although related, categories. First, at its most visible and impressive in the form of and as the effects of great industrial systems, communication was conceived of as part of the growing national economy and technological society.[15] In this context, it was considered first and foremost a resource and a target for development, being largely of interest only insofar as it was seen in the light of the functioning of an industrial unit. Nevertheless, this functionalism was not restricted to the industrial sphere, even

13 Harlow, *Old Wires*, 341.

14 David Noble, *America by Design: Science, Technology, and the Rise of Corporate Capitalism* (New York: Alfred A. Knopf, 1977), 6. Vilfredo Pareto, an Italian sociologist and economist, is said to have been the originator of the concept of the "social system" (Daniel Wren, *The History of Management Thought* (5th edn), (Hoboken: John Wiley & Sons, 2005), 199).

15 e.g., *Recent Economic Changes in the United States*, Vol. I (1929), 91.

though it drew its inspiration largely from it, and came to be regarded as the foundation for *any* administrative task. It was in this period that communication, representing the modern technological society, was linked irrevocably with the general need to control, whether in private or public organizations — a link that is still highly effective. It was precisely this control aspect that gave telephone technology its "outstanding competitive advantage over the older agencies of point to point communication," as an official report put it in 1933.[16]

Second, given the power of communication in early-nineteenth-century society, there emerged an interest in systematizing the whole communication process and unveiling its hidden mechanism, its very "logic." The systematization of communication emanated, of course, from the same movement in which industrial processes became rationalized according to a general scheme for making them function more efficiently, and was thus intertwined with the institutionalization of scientific research. While actual telecommunication technologies and the systems they enabled did largely constitute the environment wherein communication originally was brought out as a social question, thinking about communication was articulated more in technical terms, the more these technologies were incorporated into a sphere of scientific research. The concept of communication, while the motivation for its systematic examination stemmed from a particular historical stage wherein its instrumental positivity had become evident, was materialized within frameworks that presented themselves as objective, and dispensed with all its "subjective" and thus irrational connections in order to capture its innermost "essence." It is in this environment that the "scientific" conception of communication was first worked out, preparing the ground for subsequent generalized models of communication systems deeply rooted in academic treatises. These lineages form the path that we must follow in order to trace the formations laid down for and in the course of communication, and to investigate the role communication has been given in modern American society.

Communication and Law

By the time the telephone system was established it had become clear that the influence of the new communications and transportation technology, coupled with machine industry, on the transformation of the social conditions of American society was fundamental. Specifically, the evolution of the telephone system along

16 Malcolm Willey and Stuart Rice, *Communication Agencies and Social Life* (New York: McGraw-Hill Book Co., 1933), 148. On the spread of the telephone and the relating technical expertise in the United States, and the difficulties and contestations surrounding that, see Ithiel de Sola Pool, ed., *The Social Impact of the Telephone* (Cambridge: MIT Press, 1977); Pool, *Forecasting*; Marvin, *Old Technologies*; Claude Fischer, *America Calling: A Social History of the Telephone to 1940* (Berkeley: University of California Press, 1992). By 1912 the telephone had outstripped the telegraph by indisputable numbers: at that time it represented 91.8 percent of the total miles of single wire used by both the telegraph and the telephone; 83.1 percent of the number of employees; 79.4 percent of salaries and wages paid; 81.4 percent of capital stock and outstanding bonds; and 79.8 percent of income ("Growth of the Telephone," *Wall Street Journal*, (June 17, 1914)). For an account of the social and cultural aspects of the use of the telephone at the end of the twentieth century, see James Katz, *Connections: Social and Cultural Studies of the Telephone in American Life* (New Brunswick: Transaction, 1999).

with other communication technologies was an integral part of the transformation of the era's "information regime,"[17] which also explains why this system could be presented as a new social force.[18] Thus communication in its modern forms was not only conducive to the dissolution of the shared structure of meaning and the undermining of the laissez-faire order of mercantile and small-producer capitalism: it was one of the most important causes of this process.[19] The rearticulation of social conditions to this extent clearly would not have been possible without the improvement of communication systems and the establishment of their social status. Sociologist Charles Horton Cooley, for example, writing in the early nineteenth century, depicted the social transformation brought about by the railroads, the telegraph, newspapers, and the telephone as revolutionary in every aspect of life.[20] Therefore the improvement of means of communication cannot be thought of as separable, nor even as distinguishable, from social development in general. Communication as the basis of society naturally always builds this connection, but especially in developed industrial societies, social events are organized expressly as communicative events: as events in a communication environment in and through which social relations are established as social and the common experience of what is going on comes in the sphere of shared meanings.

Modern communication systems were not merely vehicles of change; by constituting the field wherein this change could be taken as an object of speech they formed the preconditions of the thinking of the social reorganization itself. This idea was articulated in contemporary writing mostly as a notion of communication as the factual site of the political. It was also the conclusion to which the progressive historians Charles and William Beard, for instance, came in 1930, for according to them the improved means and forms of communication had created an "intellectual climate in which governments make and execute policies."[21] Though communication technology was still seen strictly as a means to distribute information, as opposed to a truly constitutive power, the new systems nevertheless were conceived as having become established not only as the essential environment for political action but also as its necessary precondition.

17 Bruce Bimber, *Information and American Democracy: Technology in the Evolution of Political Power* (Cambridge: Cambridge University Press, 2003), 74; see also W. Bernard Carlson, "The Telephone as Political Instrument: Gardiner Hubbard and the Formation of the Middle Class in America, 1875–1880" in *Technologies of Power: Essays in Honor of Thomas Parke Hughes and Agatha Chipley Hughes*, ed. Michael Allen and Gabrielle Hecht (Cambridge: MIT Press, 2001).

18 Robertson Barrett, "The Telephone as a Social Force," *Bell Telephone Quarterly* 19, (2), (1940).

19 e.g., Robert Park and Ernest Burgess, *Introduction to the Science of Sociology* (Chicago: University of Chicago Press, 1924 [1921]), 558–9. This formulation should not conceal the fact that telecommunication systems, consisting of innumerable chains of interaction among individuals, machines, practices, codes, rules, formations of knowledge, institutions, and cultural and economic climate, are not considered here as an "agent" or the only cause of social change in general. Rather, by focusing on the historical conditions and configurations of communications systems it becomes possible to investigate this change within a horizon in which the relationship between communication and telecommunication becomes meaningful.

20 Charles Cooley, *Social Organization: A Study of the Larger Mind* (New York: Charles Scribner's Sons, 1912). See also Warren Susman, *Culture as History: The Transformation of American Society in the Twentieth Century* (New York: Pantheon Books, 1984), 252–70; Quandt, *Small Town*.

21 Charles Beard and William Beard, *The American Leviathan: The Republic in the Machine Age* (London: Jonathan Cape, 1930), 6.

The telephone system, "the greatest industrial and commercial achievement of the American people,"[22] regarded still in terms of an organism, constituted a whole new mode of publicity. Arnold Bennett, a British novelist writing in *Harper's Monthly*, understood that the telephone network, in connecting innumerable private households and businesses together, engendered a new field of communicativity. "Just as I think of the big cities as agglomerations pierced everywhere by elevator-shafts full of movement," he said, "so I think of [the telephone lines] as being threaded under pavements and over roofs and between floors and ceilings and between walls, by millions upon millions of live filaments that unite all the privacies of the organism — and destroy them in order to make one immense publicity."[23] As it was hoped the telegraph would bring people together, so the telephone seemed to be capable of establishing a unifying interconnectedness by means of speech, becoming a "national speech organism."[24] Thus, a contemporary commentator characterized the social potential of the telephone in 1913 as follows:

> Behold a highly developed and well organized system of communication touching human activity at every point — an essential factor in the social and industrial life of the nation — cities, towns, villages and hamlets, though widely separated, farms, factories and firesides everywhere intimately bound together by countless avenues of speech. [. . .] The incessant exchange of information and ideas which this system makes possible, puts individuals and localities on common ground — sweeps away sectional prejudices — advances civilization. [. . .] While the telephone, together with other methods of communication — the railway, the mail and the telegraph — have been the means by which the occupation and civilization of this great country have been accomplished, the telephone has provided a facility peculiar to itself, in that, over areas of great and small, it has made it possible, by spoken word, to bring together — mind to mind — the vast majority of the American people.[25]

The belief in a novel kind of communicative space, to be fulfilled with the aid of the telephone and newspaper, struck the social imagination. Although the preconditions for close national unity had clearly mostly faded, it was still necessary to envision society as an intelligible political and cultural whole. In fact, the yearning for moral unity, in particular, gained new momentum as many progressive thinkers hoped that advances in communications technology would result in mutual

22 Burton Hendrick, "Telephones for the Millions," *McClure's Magazine* 44, (1914): 45.

23 Cited in Peter Young, *Person to Person: The International Impact of the Telephone* (Cambridge: Granta Editions, 1991), 81.

24 Merrill Denison, *Advancing America: The Drama of Transportation and Communication* (New York: Dodd, Mead & Co., 1936), 229. Herbert Casson described the telephone as "an organ of the social body" (Herbert Casson, "The Social Value of the Telephone," *Independent*, (October 26, 1911): 899).

25 Union Bethell, *Telephone Engineer* 9, (4), (1913): 184. See also Herbert Casson, "The Future of the Telephone," *World's Work* 19 (May 1910): 12916, 12918. In a similar vein, Robertson Barrett stated about three decades later that "[o]ne of the most significant of the telephone's contributions to the life of our nation, as a nation, has been that it has helped to bring about what may truly be called a sense of national solidarity" (Barrett, "Social Force," 131). On unification and integration seen as an outcome of a technological imperative in telecommunications, see Dallas Smythe, *The Structure and Policy of Electronic Communication* (Urbana: University of Illinois, 1957).

identification and common values — that is, the characteristics of a community in the strict sense — to match the increasing physical and economic integration. Charles Cooley, John Dewey, Robert Park, and some other progressives were among the first thinkers of industrial America to understand the community-creating power of communication, and to take it as an object of analysis.[26] Communication built commonality: it was not only about generating communication but also, more fundamentally, about generating through communication.[27] At the same time, the technological systems that organized communication as part of the industrial structure of society increased in importance, as they seemed to provide the answer to the problem of community.[28] If the telegraph did not actually

26 See Charles Cooley, "The Theory of Transportation," *Publications of the American Economic Association* 9, (3), (1894); Charles Cooley, "The Process of Social Change," *Political Science Quarterly* 12, (1), (1897); Charles Cooley, *Human Nature and the Social Order* (New York: Scribner, 1902); Charles Cooley, *Social Organization*; Charles Cooley, *Sociological Theory and Social Research* (New York: Henry Holt & Co., 1930); Charles Cooley, *On Self and Social Organization*, ed. Hans-Joachim Schubert (Chicago: University of Chicago Press, 1998); John Dewey, "Ethics and Physical Science," *Andower Review* 7, (42), (1887); John Dewey, "The School as Social Center," *Proceedings of the National Education Association* (1902); John Dewey, *Democracy and Education: An Introduction to the Philosophy of Education* (New York: Macmillan, 1917); John Dewey, *The Public and Its Problems: An Essay in Political Inquiry* (Chicago: Gateway Books, 1946 [1927]); Park and Burgess, *Introduction*, (1924 [1921]); Robert Park, *Society* (Glencoe, IL: Free Press, 1955); Robert Park, *On Social Control and Collective Behavior*, ed. Ralph Turner (Chicago: Phoenix Books, University of Chicago Press, 1967); Robert Park, *The Crowd and the Public and Other Essays*, ed. Henry Elsner Jr., (Chicago: University of Chicago Press, 1972). Commentaries, especially from the communicative perspective, include James Campbell, *Understanding John Dewey: Nature and Cooperative Intelligence* (Chicago: Open Court, 1995); David Fott, *John Dewey: America's Philosopher of Democracy* (Lanham: Rowman & Littlefield, 1998); Ellsworth Fuhrman, *The Sociology of Knowledge in America, 1883–1915* (Charlottesville: University Press of Virginia, 1980); Larry Hickman, *John Dewey's Pragmatic Technology* (Bloomington: Indiana University Press, 1990); Larry Hickman, *Philosophical Tools for Technological Culture: Putting Pragmatism to Work* (Bloomington: Indiana University Press, 2001); Lester Kurtz, *Evaluating Chicago Sociology* (Chicago: University of Chicago Press, 1984); Fred Matthews, *Quest for an American Sociology: Robert E. Park and the Chicago School* (Montreal: McGill-Queen's University Press, 1977); Steven Rockefeller, *John Dewey: Religious Faith and Democratic Humanism* (New York: Columbia University Press, 1991); Robert Westbrook, *John Dewey and American Democracy* (Ithaca: Cornell University Press, 1991).

27 It was especially Dewey who understood communication in more ontological terms as the precondition of a community: he conceived community fundamentally as communication, existing *in* communication, not just *by* communication. According to him, "[t]here is more than a verbal tie between the words common, community, and communication. Men live in a community in virtue of the things which they have in common; and communication is the way in which they come to possess things in common" (Dewey, *Democracy and Education*, 5). However, while promoting communication technologies in the creation of a truly political community, Dewey also believed in objective inquiry, commonly identifiable values, and rational social management. Besides Dewey, a number of voices speaking up on behalf of democratic governance, as opposed to social control, were articulated during the first half of the twentieth century. For instance, Harold Laski of the London School of Economics and Frank Knight of the University of Chicago resisted the application of science to society in order to facilitate control, yet they failed to reorient the sentiment that was animated by the possibility of scientific social control (John Jordan, *Machine-Age Ideology: Social Engineering and American Liberalism, 1911–1939* (Chapel Hill: University of North Carolina Press, 1994), 157–65).

28 Seeing technology, especially communications technology, as the solution to the community problem put thinkers such as John Dewey and Charles Cooley among those who were enchanted with mechanization and efficiency, and who sought, in the final analysis, technological answers to social and political questions (Quandt, *Small Town*, 66; Carey, "Cultural Approach," 10. See also James Startt and William Sloan, eds., *The Significance of the Media in American History* (Northport: Vision Press, 1994), especially chapters 1 and 16; Daniel Czitrom, *Media and the American Mind: From Morse to McLuhan*

succeed in establishing a harmonious community, the far more efficient telephone, which inherited the hopes invested in the earlier communication system, seemed to be able to fulfill expectations and "bind this country together with a constantly changing but ever-present web of words."[29] The telephone was no doubt regarded as an important means of engendering and organizing the society into a coherent system prior to national broadcasting.[30] This view was reflected, for instance, in the address by American Telephone and Telegraph Company (AT&T) vice president Nathan C. Kingsbury before the Central States Conference on Rail and Water Transportation in 1917. "In the history of the telephone," he stated, "there have [. . .] been no greater achievements than the development of transcontinental telephony and of wireless telephony. [. . .] This is a final blow to sectionalism. The East is no longer separated from the West, nor the North from the South. [. . .] Co-operation is rendered easy, natural, necessary, permanent."[31]

It seemed apparent that communication resided in the core of the modernization process that permeated the whole society, and, in the final analysis, it was nothing other than the expression of the dynamic unfolding of society itself. Of course, it must be noted that communication was not — with some notable exceptions such as for instance Charles Cooley and John Dewey — articulated as an constitutive question concerning the realm of communicativity but was stuck in the instrumental conception: communication was something by which one could create "intellectual climates" rather than being related to the emergent way of being that made this creation possible in the first place.[32] Even the exceptions to the rule manifested the thinking about community in terms of this instrumentality. Yet, communication began to carry new weight in the social imagination as an integrating force that it was hoped would create a mutual understanding, shared values and a common purpose.

Thus the renewal of communication in a new technical mode appeared as a two-pronged question. On the one hand proliferation of connections speeded up the general economy and increased the complexity and fluidity of social

(Chapel Hill: University of North Carolina Press, 1982); Ronald Tobey, *The American Ideology of National Science, 1919–1930* (Pittsburgh: University of Pittsburgh Press, 1971), 176).

29 Page, "Social Aspects" (in *Modern Communication*), 24. See also "Telephone vs. Telegraph" (December 10, 1904) in Vinton Sears, ed., *Telephone Development: Status of the Industry, Scope and Effect of Competition* (Boston: Barta Press, 1905), 90. The slogan of the Bell System from 1907 to 1914 — "one system, one policy, universal service" — could also be seen as implying a belief in the capacity of an ubiquitous communications network to contribute to national unity and equality of opportunity (Milton Mueller Jr., *Universal Service: Competition, Interconnection, and Monopoly in the Making of the American Telephone System* (Cambridge: MIT Press, 1997), 5, 92, 96–101).

30 Pool, *Forecasting*, 71.

31 N. C. Kingsbury, "Co-operation and the Telephone," *Telephone Engineer* 17, (3), (March 1917): 146. Kingsbury delivered another speech containing almost identical phrasing at the opening of long-distance telephone communication between the Atlantic and the Pacific in 1915 (see Pound, *Telephone Idea*, 24–5). For a survey on the long distance communications, see "Communicating Over Great Distances," *Scientific American*, (June 5, 1915).

32 Thus, the modern tools of communication "give man the power to do many things with greater facility," stated a commentator in 1932, "but they do not control the degree nor the direction in which he uses that power" (Page, "Social Aspects" (in *Bell Telephone Quarterly*): 175; also in Page, "Social Aspects" (in *Modern Communication*), 20–1).

and industrial processes. Thus it multiplied the "burdens of government"[33] and enhanced the need for organization and planning. The telegraph alone had grown and accelerated collective forging of social relations to the extent that it, in the main, disintegrated the very order on the basis of which it had developed, and the telephone increased many times the level of the complexity and the rate of social and communicative processes.[34] On the other hand, communication, especially in the form of developed communication technologies, could be seen at the same time to be the central tool for resolution of this problem. This was because communication systems such as the telephone enabled better control by increasing the efficiency of coordination.[35] Furthermore, telephone systems constituted in themselves a giant industry the network structure of which provided a recognizable representation for grasping the new conditions of societal activity in which distant actions seemed to have immediate and perceptible effects. Therefore they could be viewed as models for making the governance of society more effective and businesslike, which was what the contemporary critique desired in opposition to the "corruption" of the government.

After the telephone and the immediate interpersonal communication it set up was socially established, it emerged as an indispensable resource with respect to the governance of numerous activities, becoming an "absolute business and social necessity," as it was described in 1920.[36] This was manifested both in the increase in telephones and telephone lines and in the growing use of them on a routine basis. So, the telephone was no longer just an auxiliary device in social practices but was instead taken for granted as the precondition of these practices. "Organized society," stated a contemporary, "cannot exist without inter-communication, and the quicker and more certain it is, the greater is the uplift of humanity in the intimate relations of social units."[37] The telephone was a definite foundation not only of the corporate domain but also, increasingly, of the functioning of the society itself, the principles of which were taken under intensive examination.

33 Beard and Beard, *American Leviathan*, 6.

34 See Roger Burlingame, *Engines of Democracy: Inventions and Society in Mature America* (New York: Charles Scribner's Sons, 1940), 130.

35 On the technique in general, being both the creator of the areas to be organized and the tool by which this task of organizing could be carried out, see George Soule, *A Planned Society* (New York: Macmillan, 1932), 269. For a brilliant study on mass communication as both Satan and Savior in progressive social thought, see John Peters, "Satan and Savior: Mass Communication in Progressive Thought," *Critical Studies in Mass Communication* 6 (1989): 247–63.

36 Frank Bohn, "The Association and Its Needs," *Telephone Engineer* 24, (4), (October 1920): 20; see also Chandler Murphy, "Regulation of Service," *Telephone Engineer* 23, (4), (1920): 17. Belief in the inevitability of the telephone was rife throughout the period, and was in evidence from "The Telephone in Modern Business," *Telephony* (November 1901), through "The Indispensable Telephone," *Telephony* 12, (5), (November 1906), and furthermore to a citation from the *Sarasota Herald* in October 1934: "The telephone [. . .] has become such a necessary part of business and social life that it is impossible to see how the people were able to get along prior to its invention" (cited in Dennis Cooper and E. L. Harris, *The People Machine: An Illustrated History of the Telephone on the Center West Coast of Florida* (3rd edn). (Tampa: GTE Florida, 1988 [1971]), 104).

37 Murphy, "Regulation," 17. This was naturally realized earlier. Thus S. P. Grace could write in the early twentieth century: "[s]hould [. . .] a telephoneless condition ever eventuate [. . .] the whole nature of public and private life would be totally changed in a miraculously short space of time" (S. P. Grace, "Common Battery for Small Exchanges," *Telephony* 4, (2), (1902): 95).

Lester Frank Ward was the first person in American social science to articulate the notion that history was subject to scientific control and direction, and would have had sociology mobilized into systematic disclosure of the rules according to which social laws could be controlled.[38] For Ward, distancing himself from the static social Darwinism of William Sumner, movements within society appeared not as preordained for, although still following natural laws, they were now seen as governable in just the same way as natural forces have proved rationally governable thanks to the natural sciences. Here, the model of natural sciences provided a way for Ward to rethink the status of social laws, because if the same laws governed both physical world and human society, discovery of these laws would allow rational control. This let him conceive politics as the governance of these laws "just as water, wind, and electricity are controlled."[39] In this way, politics as the sphere in which social relations were articulated was to be supplanted by the "scientific" control of social laws. Thus, communication, the voice and mark of the social, became something that could be subjected to communal regulation, because the laws that directed its course could be formulated not only for analysis but also for manipulation.

In this respect Ward paved the way for technocratic thinking by articulating a program to capture the inherent "essence" of society in the form of social laws and to control these laws in a systematic manner. Politics, in this program, came to be replaced by a set of supposedly uncontroversial and objective rules, based on scientific analysis.[40] Insofar as for Ward the possibility of directing the development of history was important, it was this very aspect that was removed from discourses of social engineering proper. In those, the instrumental dimension of effective social control excluded questions concerning common objectives and the direction of history — in other words, the whole possibility of politics. Nevertheless, Ward demonstrated how easily the concept of controllability of the social was translated into a question of technology. This was not, however, just accidental, for the increasing awareness of the general significance of instrumental efficiency constituted to a great extent the conceptual context within which early-twentieth-century American society and its future were posed as objects of thought.[41]

This notion, which substituted Sumner's ineluctable natural development by scientifically generated progress, and competition, as manifested in the survival of the fittest by natural selection, by mutual cooperation and rational negotiation, opened up a new relation to thinking about communication as well. For insofar as

38 Hanson, *Democratic Imagination*, 208–10; Ross, *Origins*, 91; Hofstadter, *Social Darwinism*, 68–84.

39 Lester Ward, "The Claims of Political Science," in Lester Ward, *Glimpses of the Cosmos*, 6 vols. (New York: Putnam, 1913–18 [1884]), 3: 334. Here, Ward drew on the positivism of Auguste Comte, according to which technocratic manipulation, based on the science of society, could be extended to social forces (Gillis Harp, *Positivist Republic: Auguste Comte and the Reconstruction of American Liberalism, 1865–1920* (University Park: Pennsylvania State University Press, 1995), 19).

40 Lester Ward, *Dynamic Sociology*, 2 vols. (New York: Johnson Reprint, 1968 [1883]), 1: 16, 60; 2: 205–10.

41 e.g., Samuel Haber, *Efficiency and Uplift: Scientific Management in the Progressive Era, 1890–1920* (Chicago: University of Chicago Press, 1964); Martha Banta, *Taylored Lives: Narrative Productions in the Age of Taylor, Veblen, and Ford* (Chicago: University of Chicago Press, 1993).

they appeared to enable the general economy of social interaction, the conditions of communication were translated from natural laws external to human authority to principles pertaining to the inherent nature of human society itself. Although still remaining within the sphere of natural law, these principles were now included as workable elements of the human realm. In this way the practical, instrumental viewpoint lent itself naturally to dealing with social questions. So, communication, a sign and signal of the social, became subordinated to social direction and regulation by virtue of the fact that the underlying laws that determined its course could be used for manipulation by sociological investigation. The forms of multifarious communicative relations appeared in the perspective of this applied, "dynamic sociology" no longer as reflecting an unchangeable blueprint, but were seen instead as following an intelligible and, eventually, controllable law.

What appeared as the most crucial question, as opened up and addressed by the reform Darwinism that Ward systematically articulated, was not so much concerned with historicity, but controllability of the social domain: instead of considering the social as being realized in the course of its becoming articulated, it drew attention mainly to methods for the proper organization of social life. With the new ideas about "social control," as put forward by the sociologist Edward A. Ross, Ward's nephew-in-law, and other organic social scientists, the need to modify individual conduct to conform to the public interest was established as a pivotal concern in social thought.[42] The aim of Ross's pioneering work in this field, as he saw it, was to "enable society to go about the business of control in a scientific way."[43] A technocratic conception of social control according to which control commonly referred to a social process rather than an exercise of political will, focusing on the purportedly objective principles of efficient managerial functioning, began to take hold and to direct the imagination in the writings of the early sociologists and the reformist intellectuals of the period.[44]

The step in this direction, in terms of social theory, by which politics was replaced by technique, was legitimized by "scientific method," which established the domain within which controllability could be carried out. This method, being essentially "analytical and rational," was viewed as opposed to the "instinctive, emotional, rule-of-thumb operations of historic politics."[45] In addition, it leads

42 For an interesting account of the relationships among concepts of state, democracy, and social control, see Melossi, *Social Control.*

43 Edward Ross, *Social Control* (New York: Macmillan, 1901), 436f1. However, it should be kept in mind that as far as the progressive thinkers writing on this theme were concerned, social control referred not to physical coercion but rather to inner control enforced through shared ideas and public opinion (see Eldon Eisenach, *The Lost Promise of Progressivism* (Lawrence: University Press of Kansas, 1994), 76, 87, 89).

44 Ross, "Development," 108, 127–30; Ross, *Origins*, 230–33, 249–52, 311, 388; Dorothy Ross, "Modernist Social Science in the Land of the New/Old" in *Modernist Impulses in The Human Sciences, 1870–1930*, ed. Dorothy Ross (Baltimore: Johns Hopkins University Press, 1994), 171; Schwendinger and Schwendinger, *Sociologists of the Chair*, 109, 127, 156, 203–16; Sklansky, *Soul's Economy*; 201–4. In a similar vein, George Herbert Mead argued that "[w]hat society is struggling to accomplish is to bring [the] social side of our conduct out so that it may [. . .] become the element of control" (George Herbert Mead, "Review of C. L. Morgan: An Introduction to Comparative Psychology," *Psychological Review* 2 (July 1895): 401).

45 Beard and Beard, *American Leviathan*, 8. Walter Lippman held already in 1914 that science would save democracy by giving impartial solutions to all economic, political and social problems (Walter

unavoidably to the right decisions relative to the situation: "[I]t calls for the collection of pertinent facts, the formulation of conclusions on the basis of facts, and the execution of policies in accordance with the requirements of the fact situation."[46] So, through the coalescence of Darwin's conception of evolution with scientific method, a possibility arose for a new relationship to the social. The far-reaching significance of this cannot be underestimated, for this is why a whole sphere of practices became firmly entrenched on a foundation of law — in the sense of predetermined relationships and unchanging, systematically applicable procedures. This was related to the fact that the possibility of control was only conceivable by virtue of the scientific edifice that was itself grounded in the same principle of controllability. At the same time, administration — linked with ideologically nonpartisan rationality and objective methods — was separated from politics — which was linked with corruption, incompetent elected officials, and special interests.[47] Thus, whereas republican individualism drew a line between society and politics, now the line was between politics and administration.

As the changed circumstances demanded a rational recasting of society, a mechanism for reform was thus discovered not in the organic model but rather in the alliance between evolutionary theories and the repeatable method of science that posed the conditions of communication as a factual structure.[48] In this way a still extant project was initiated, in the course of which these conditions showed up as able to be understood in consistent and quantifiable categories. At the same time they were enshrined in the system of metaphysical science as something conceptually uniform and numerically definite. It is this project — although heterogeneous as well as discontinuous — wherein communication assumed a growing "beingness" in terms of the forms in which it was introduced as an object of scientific analysis.[49] Of course, there is nothing accidental in this, for science

Lippman, *Drift and Mastery* (Englewood Cliffs: Prentice-Hall, 1961 [1914]), 151, 154, 165. See also Carroll Pursell, *The Machine in America: A Social History of Technology* (Baltimore: Johns Hopkins University Press, 1995), 204; Tobey, *American Ideology*, 3–18, 167–77, 217–27).

46 Beard and Beard, *American Leviathan*, 8.

47 William Akin, *Technocracy and The American Dream: The Technocrat Movement, 1900–1941* (Berkeley: University of California Press, 1977), 2–5, 112; Tobey, *American Ideology*, 15–16; Hanson, *Democratic Imagination*, 244–5; Wiebe, *Search of Order*, 185. Of course, the idea of the separation of politics from administration was already in the air by the end of the nineteenth century, thanks especially to Professor Frank J. Goodnow who gave it a systematic formulation in his book *Politics and Administration: A Study in Government* (New York: Macmillan, 1900). Yet, attempts to dispense with the ultimate conflict and uncertainty that is part of the political realm by turning to an administrative outlook became the hallmark of the progressive era (see Haber, *Efficiency and Uplift*, 103–4; Lustig, *Corporate Liberalism*, 152–4).

48 For the displacement of the organic model and the idea of origins according to the conception of evolution and growth in American constitutional theory, see Kahn, *Legitimacy*, 66–96.

49 To what extent can we speak of the transformation of communication per se? As already pointed out, I have rejected the option of conceptualizing its development as a unified history, for it is in the process of a communication unveiling its own history alone that frameworks are both erected and dismantled through which its "character" can be addressed as an object of thought. However, there is something commonly recognizable in the discussion about "communication," although the fields within which it has been given a meaningful interpretation have mutated through different formations. Furthermore, the question is not of the juxtaposition of eventfulness and beingness (an event can be as being-like as a being eventful), but rather of the types of form through which communication was expressed within a sphere of systematic thinking.

isolates its objects and thus causes, to a large degree, being-like existence to be tested, measured, and assessed. Therefore it is not surprising that communication adopted an inherent structure and mechanism of its own out of the indeterminate oscillation wherein it both conditions and is being incessantly conditioned. Hence, what has defied a comprehensive definition — namely, the always-elusive "nature" of communication — started to be considered as a positive term.[50] Yet the positivity introduced here is always its own positivity, not an external quality attributed to it. What the emerging figure of communication was suggestive of, in a way, was society's own image in that the scientific practices that evoked it became an ever more integral part of a society that was constituted, in the final analysis, from communication.

Thus, the instrumental and conceptual means provided by communication systems were given buoyancy by "scientific method," as successfully pursued in the natural sciences, in the light of which the full potential of communication was to be realized within a rational guidance system. The nature of this method consisted in a number of things. First, it separated the incidental and insignificant from what was essential and necessary, and in so doing forged a relationship to truth. Second, it replaced contingent, subjective opinion by a rational and objective procedure that did not leave room for individual interests or preferences. Third, this procedure had a coherent, endlessly repeatable nature, which did not change with time and context, whereas its objects, in turn, appeared independent from the method, which only revealed them without having any influence on their existence as such. Lastly, the method was insensitive to its objects: it could be applied in principle to any object regardless of their particular type or distribution.

This method constituted the foundation for scientific control in the social sphere, but also laid the basis for communication becoming a focus of systematic production of knowledge. From the technical point of view, the latter virtually developed hand in hand with the former, as communication in the form of telecommunication required dependable and effective transmission systems that both necessitated scientific research and were an outcome of it. Scientific interest in communication, including the emerging field of communications research, strengthened concurrently with the improvement of technologies of communication. It was connected to the growing sentiment, supported by the tradition of American functionalism, that communication as a social process, after having been analyzed and formalized, could be used to maintain and steer the social

50 Communication, here, is not understood as an imaginary totality estranged from its own being. What is at issue here is tracing the line of thought through which communication has become understood as a rational system, not implying that communication would or could be reduced to a rational scheme "in itself," as the conception of communication in itself is only possible in these or similar schemes. However, clearly it is predominantly by way of these kinds of discursive frameworks that communication is exposed to us as an object of systematic thinking, which makes them culturally, if not theoretically, the foremost loci for the investigation of the administrative history of communication.

system.[51] Thus clarification of the communication process was not merely an academic but also a thoroughly political question.

Despite the fact that the scientific control of society was presented as a collective mission, the actual determination of social processes, as it were, did not depend primarily on conscious decisions. Communication was conducive to increasing the level of social ordering by engendering and differentiating positions, practices, and connections prior to and alongside discourses on rationalization and social planning. It routinized forms of interaction and specified operational chains of action, allowing the reformulation of tasks and necessitating rearrangement of organizational structures. Through creating classes of action and types of organization, it was also involved with the differentiation, specialization, and serialization that came to mark the industrial society. It is in this connection — that is, in the conjunction of communication and knowledge — where singular communicative relations became inscribed into the body of the scientific edifice. In this respect, the growing organization of society was indissociable from the inquiry into the nature of communication.

Thanks to the historical linkage between communication and a general method or law, communication was considered as capable of being brought within the sphere of conscious command, which reflected the ambition and hope of society taking control of its own destiny. While communication, in the form of social relations and processes, was wanted under rational control, it was, however, this same communication that would perform the control required. Seen from this perspective, concern about communication did not just reflect the need to direct the course of society and thus to take responsibility for its own being, for it was communication that established the space and defined the ways within and by which this concern could be expressed: if communication was a means, it was a means for society to exist and maintain itself as a society.

The concept of society based on the idea of a founding law as a mode of social governance was gradually given explicit formulation through the view of society as an interacting whole of different social groups, which replaced the old idea of it as an aggregate of atomistic individuals. Now the basic unit of society was the activity of groups rather than of an individual, and it was invested with concepts such as "group reason" and "the group mind."[52] This collectivist turn implied an ontological shift from the conception of the individual as an independent and self-sufficient entity to one in which he or she is defined in and through the broader rationales and interrelated processes of society. It, in turn, was a fundamental part of the process that prepared the ground for a more positive view of the state and for national planning evolving as a major issue of American public policy.[53]

51 Hanno Hardt, *Critical Communication Studies: Communication, History and Theory in America* (London: Routledge, 1992), 14.

52 e.g., Arthur Bentley, *The Process of Government: A Study of Social Pressures* (Evanston: Principia Press of Illinois, 1935 [1908]), 204, 206. See also Schwendinger and Schwendinger, *Sociologists of the Chair*, 123, 193–4; Lustig, *Corporate Liberalism*, 110–29.

53 Akin, *Technocracy*, 1–2; David Wilson, *The National Planning Idea in U.S. Public Policy: Five Alternative Approaches* (Boulder: Westview Press, 1980), 26; Lustig, *Corporate Liberalism*, 39; Quandt, *Small*

Progressive thinkers like Charles Cooley, Herbert Croly, Walter Lippmann, and John Dewey, attempting to formulate the outlines of a new, industrial liberalism, agreed that the supposed natural harmony of economic competition was not capable any longer of maintaining social stability, which depended instead on the idea of cooperation. In this context, the exposition in the second half of the nineteenth century of communication in terms of competition was rejected.

It is easy to read from the contemporary discourses that during the time of the telegraph it was communication that was the integrating social power: it united the community, while political economy, in turn, defined communication as a universal competition among atomistic individuals pursuing their own egoistic self-interest. Now, however, communication assumed a novel character because the idea of competition did not correspond any longer to experience of the present situation or the requirements it placed on social action. The great industrial systems that were seen as epitomizing the intrinsic nature of modern technical society established the frame of reference within which communication came to be considered. As the central processes of society were depicted as systemic, and as it was understood that what characterized the society as such was no longer competition but cooperation and the aspiration for smooth and efficient action, the concept of a system became the reference point through which the progressive idea of the use of scientific methods in administration was entwined with the aim of efficiency.[54] In contrast to the European tradition, social governance was not organized as a comprehensive program in American political thinking. Yet insofar as the political tradition marked by localism and fragmentation became increasingly restructured on the basis of law and a general method,[55] it was impossible to disentangle communication as a systemic instance from the realization of this law.

The rise of communication as a theoretical and administrative concern must thus be seen against this background. During this period communication became conceived of as a distinct and determinable object of organized discourse. At the political level, the advent of communication as a separate administrative category was consolidated when interstate telephone and telegraph were included under the regulation of Interstate Commerce Commission in 1910,[56] which marked the

Town, 23–35, 61. However, as is known, all this changed during the Cold War period as rational-choice theory, based on competitive individualism, gained ground in social thought in opposition to theories supporting communism and collectivism.

54 Karl, *Executive Reorganization*, 67; Tobey, *American Ideology*, 3–18, 32–61, 133–59, 167–98; Pursell, *Machine in America*, 204. Progressive thinkers continued to believe in the essential harmony between groups in the industrial society. It was the perspective of unity rather than conflict that fired their imaginations (see Lustig, *Corporate Liberalism*, 136–42).

55 As Christopher Dandeker has noted, the motivation to create a consistent national power structure grounded on law inhered largely in the special arrangements caused by war and civil and industrial unrest (see Dandeker, *Surveillance*, 124).

56 The Mann–Elkins Act of 1910, as part of the progressive reforms, extended the authority of the Interstate Commerce Commission (ICC) and made telephone and telegraph carriers subject to its regulation. The ICC, created by the Interstate Commerce Act, had the authority to regulate telephone companies as common carriers and interstate telephone services between 1910 and 1934, when this authority was transformed to the newly founded Federal Communications Commission. However, the ICC is considered

moment when telecommunications came under national control and became an established category of national policy. Notwithstanding the extension of federal regulation to include telephone and telegraph services, however, the new legislation was not originally focused on communications: the main interest remained in the railroads and the need for specialized jurisdiction to review the orders of the Commission.[57] Nevertheless, the main point in this development should be pointed out: the very idea of neutral regulatory agencies, now applied to communications within an organized legislative structure, reflected the progressive belief in the possibility of objective social knowledge and the isolation of administration in an independent professional bureaucracy.[58]

At the theoretical level, one of the first definitions that ascribed a discrete conceptual identity to communication, formulated in 1928 by Ivor Armstrong Richards, an English literary critic and author, stated that "communication takes place when one mind so acts upon its environment that another mind is influenced, and in that other mind an experience occurs which is like the experience in the first mind, and is caused in part by that experience."[59] Clearly aiming at universality, the definition rested on an instrumental view of communication, seeing it as an effective social relationship. Here, the concept that extracted a core, an essential substance of communication from all the processes in and within which it occurred, started to get articulated in positive terms, which largely reflected the general professionalization of social scientific disciplines in terms of the natural sciences model, narrowly understood. If communication proved an attractive object to diverse administrative and scholarly practices, it was the development of

to have contributed very little to the development of regulatory policy in the communications field; see Alan Stone, *Public Service*, 159; Glen Robinson, "Title I. The Federal Communications Act: An Essay on Origins and Regulatory Purpose," in *A Legislative History of The Communications Act of 1934*, ed. Max Paglin (New York: Oxford University Press, 1989), 5–8; Stephen Skowronek, *Building A New American State: The Expansion of National Administrative Capacities, 1877–1920* (Cambridge: Cambridge University Press, 1982), 123–61; Jeffrey Cohen, "The Telephone Problem and the Road to Telephone Regulation in the United States, 1876–1917," *Journal of Policy History* 3, (1), (1991): 63–4; Gerald Brock, *Telecommunication Policy for the Information Age: From Monopoly to Competition* (Cambridge, MA: Harvard University Press, 1994), 49. For a defense of the Bell system as a public utility under private management and ownership, see Theodor Vail, "Public Utilities and Public Policy," *Atlantic Monthly* 3, (1913); and for an interpretation of the policy shift in telecommunications, Dan Schiller, "Social Movement in Telecommunications: Rethinking the Public Service History of US Telecommunications, 1894–1919," *Telecommunications Policy* 22, (4/5), (1998): 397–408.

57 Robinson, "Communications Act," 6. An institution commonly referred to as the independent regulatory commission was a political innovation of the American society in this period (e.g., Louis Galambos and Joseph Pratt, *The Rise of the Corporate Commonwealth: U.S. Business and Public Policy in the Twentieth Century* (New York: Basic Books, 1988), 45). In the administrative law discourse, the right of the state to regulate the rates and practices of a public utility was viewed as referable to the police powers of the state, the purpose of the latter being to prescribe regulations to preserve and promote the public safety, health and morals (see John Kurtz, "State Public Service Commissions," in Ernst Freund et al., *The Growth of American Administrative Law* (St. Louis: Thomas Law Book Co., 1923), 151, 154–5; Lewis Hochheimer, *The Law of Crimes and Criminal Procedure* (2nd edn), (Baltimore: Baltimore Book Co., 1904), 54).

58 Lustig, *Corporate Liberalism*, 154; Skowronek, *New American State*, 151; Robert Horwitz, *The Irony of Regulatory Reform: The Deregulation of American Telecommunications* (New York: Oxford University Press, 1989), 9, 25–6, 47–69.

59 "Communication," in *The New Encyclopaedia Britannica* (15th edition), (1982).

new communication technologies and systems, coupled with advances in scientific research, that was the context for this rising interest.

What gathered these scientific interests and technical means of communication into one distinct issue, was the perspective of control, which was only emphasized during World War I. In fact, the birth of communication research as an academic discipline has been attributed to the study named *Propaganda Technique in the World War*, written by Harold D. Lasswell in 1927. In this work, the author construed propaganda as one of the most influential forces of the modern world: one that welds separate individuals together into an integrated mass.[60] This was connected to ideas developed by the technocratic liberals of the period, as Lasswell also maintained that democracies needed propaganda to keep the citizenry in agreement with what was in their best interests — which was best determined by groups of experts. Furthermore, the beginning of mass communication research in the 1930s — for which the *Propaganda* book prepared the way — and the break it caused with previous thinking on communication with its emphasis on empirical research, was occasioned by the administrative need for concrete social information.[61]

The political technologies of communication established, for the first time, the theoretical basis of communication viewed as a structured system. "Communication" constituted an object of ever-increasing theoretical interest and became part of the administrative structure both in business and politics. Along with the strengthening of the federal government, it was Taylorism, specifically, that paved the way for the generalization of the idea of planning at national level, with Herbert Hoover becoming the first president to develop consistent national engineering within a scientific framework.[62] Communication became as a resource in the general process of rationalization.

Taylor and Communication

Alongside the utilization of technical models for rationalizing the practices of various systems, Frederick Winslow Taylor formulated a sophisticated concept of control wherein the requirements for efficiency and methodical clarity were combined into the form of a regimented, scientifically managed organization.[63] This was encapsulated in what became known as scientific management, and

60 Harold D. Lasswell, *Propaganda Technique in the World War* (London: Kegan Paul, Trench, Trubner & Co., 1938 [1927]), 220–1.

61 For the relationship between the propaganda analysis and the new communication research in the 1940s, see Kenneth Cmiel, "On Cynicism, Evil, and the Discovery of Communication in the 1940s," *Journal of Communication* 46, (3), (Summer 1996).

62 Wilson, *National Planning*, 28–9.

63 Here, as elsewhere, it is not my concern to delve into the discussion of what was the exact contribution of a putative sole creator — in this case, Frederick Taylor — to a given theory or system and what were the roles of other people often ignored. It is my task here to investigate the conceptual displacements and the administrative practices these ideas provoked, especially from the point of view of communication, not to trace their assumed origins or to identify the right persons to be credited. Taylor suffices to identify the movement in question, although surely he was less the creator of scientific management than he claimed.

after World War I, industrial management.[64] It concretized the major focus of social reform at the local level by creating a realistic program of rationalization to be implemented in industrial organizations.[65] The concept of the stability and controllability of communication found a first promising model from engineering method because it seemed to be capable of fulfilling the vision of a scientifically governed society by serving as a means of replacing social contention with indisputable answers.[66] Scientific management synthesized engineering method, the "scientific" approach, and the previous doctrines of efficiency experts and job designers into a consistent system. It established a perspective that gathered together the theoretical formulations of organization, the idea of standardization, the physical movements of workers, and material conditions for the workplace, allowing them to be analyzed within the same conceptual framework.

It is of utmost importance from the point of view of the history of communication to take cognizance of this development, as it was in factories and workshops — in a world of cranks, clamps, pins, levers, and winches — that not only was communication formulated as a distinct object of thought but also where the formalization of communication took its first steps. This is related to the fact that at the beginning of the twentieth century it had become apparent that the possibility for scientific planning of social development was based on the factual structure of communication, not on its assumed immutable principles.[67] The "Taylor System" provided the context wherein this structure could be translated into specific laws in order to rationally direct not only industry but also the whole nation, as the method of governance was seen to be basically identical in both cases.[68] The charm of this system from the administrative point of view was naturally that it suggested the possibility of specifying the "one best way" for any social activity, and this was the reason it assumed the form of a general sociopolitical program.[69]

Although irreducible to the sphere of mechanical work alone, scientific management nevertheless operated through reorganizing work practices. It can be proposed that "the passing of the frontier"[70] forced people to seek accumulation through markets, because profit had to be made more and more through competition rather than by extending markets infinitely. Experience of limits was thus tied more and more to experience of the finiteness of market relations, and it is at this point that Taylor entered the picture by presenting a "scientific" approach to the problem of the increase in production required. Scientific management was a

64 The term "scientific management" was actually coined by the Boston lawyer Louis D. Brandeis, and Taylor accepted it only with reluctance (Wren, *Management Thought*, 138–9, 272).

65 Wilson, *National Planning*, 26.

66 Jordan, *Machine-Age*, 28, 35–6, 39.

67 Karl, *Executive Reorganization*, 60.

68 See Herbert Valentine, "Social Efficiency," *Transactions of the Efficient Society* 1 (1912), 407; Haber, *Efficiency and Uplift*, xi, 26.

69 Taylor himself put this very explicitly: "The same principles [of scientific management] can be applied with equal force to all social activities: to the management of our homes; the management of our farms; the management of the business of our tradesmen, large and small; of our churches, our philanthropic institutions, our universities, and our governmental departments" (Frederick Taylor, *The Principles of Scientific Management* (New York: W. W. Norton & Co., 1967 [1911]), 8).

70 e.g., Klose, *American Frontier*.

method for meeting the need, expressed from the last decades of the nineteenth century on, for finding a way to coordinate operations and control workers in order to increase the level of production. Against this background, Taylor can be regarded as having forged a new relationship to the experience of limit, because it is in the context of the new awareness of the limits of the market that his system has left a lasting mark.

What is, then, Taylor's relevance to thought on communication? It can be said to be at least twofold. First, he established communication as one of the means by which the manager could control work done on the shop floor, thus expressing the relationship between manager and worker in terms of communication. Although it was not the only relevant perspective within which work was conceptualized, communication came to be viewed here, for the first time, as a legitimate category in the structure of work organization. Thus the success of scientific management, according to Taylor, relied on, for one thing, "the intimate cooperation of the management with the workmen, so that they together do the work in accordance with the scientific laws which have been developed, instead of leaving the solution of each problem in the hands of the individual workman," as he pointed out in his classic book *The Principles of Scientific Management*, which was first published in 1911.[71] In practice, communication between management and workers took place mostly in writing, although telephone technology was also used for this purpose. This is what Taylor, writing about the concentration of all brainwork in one location, the superintendent's office, describes in the same book:[72] "In this office [of the superintendent] every laborer's work was planned out well in advance, and the workmen were all moved from place to place by the clerks with elaborate diagrams or maps of the yard before them, very much as chessmen are moved on a chess-board, a telephone and messenger system having been installed for this purpose."[73] Insofar as the functional whole appeared to be a communication machine, communication was an integral part of the control system that this whole depended on. It was this formal structure that systematized the functional performance of a given system. This is, at the same time, why the necessary communication systems cannot be conceived only as auxiliary instruments external to work practices but instead as what made these practices possible as an organized procedure in the first place.

The theme of an underlying law governing each work performance brings us to the second point. Taylorism articulated a perspective that enabled communication to be thought of as a regular activity controlled by a general law that could be fully revealed in each particular case by the methods of science. Taylor was firmly convinced that "laws of this kind, which apply to a large majority of

71 Taylor, *Scientific Management*, 115.

72 Frederick Taylor, *Shop Management*, in *Scientific Management: Comprising Shop Management, The Principles of Scientific Management, Testimony Before the Special House Committee*, ed. Kenneth Thompson, vol. I of *The Early Sociology of Management and Organizations* (London: Routledge, 2003), 64, 107, 109–10, 127–8.

73 Taylor, *Scientific Management*, 69; see also Robert Kanigel, *The One Best Way: Frederick Winslow Taylor and the Enigma of Efficiency* (London: Abacus, 1997), 334.

men, unquestionably exist, and when clearly defined are of great value as a guide in dealing with men."[74] These laws were not, however, imposed on action from above, as a principle foreign to it, but stemmed from the specific procedures and motions needed in a given task. So the law, gathering communicative relations and events under one conceptual whole, was fundamentally a law derived from the activity in question.

Seen from the point of view of the constitutive — that is, ontological dimension of communication — Taylorism is to be understood not only as a theory of work management but also, and perhaps more important, as a way of thinking about communication within the circumstances of modern industrial society. However unusual it might seem to make Taylor a communication theoretician, here I am examining his thinking in terms of the conditions of communication. Although he was not the only theorist to incorporate communication into a production system as a distinct category, Taylor is, however, the best representative of the rationalization efforts of the early twentieth century, the object and means of which communication increasingly became too. Second, by systematizing the diverse attempts made at the time toward a more rational, foreseeable system, the model constructed by Taylor provided a general comprehension of how social communication should be organized according to rational laws. Taylorism, based on differentiation, specialization, and repetition, grew into a general cultural paradigm beyond efficiency engineering discourses proper, becoming a foundation from which communication and community began to be thought of as a single whole. Thus it is insufficient to interpret it as only a philosophy of mechanical work, since it had a strong impact on the self-conception of a whole era.[75]

A 'Taylored' system was not simply a mechanism for producing standardized artifacts; it also introduced identity and consistency into the system in question by excluding all that did not belong to it. Typically, hierarchical systems depend on a principle of exclusion. They establish an interior that abolishes and subverts something that can be interpreted as politics, aiming at substituting law for discussion. Yet, at the same time, the other — what is left out, what is abolished and displaced — always remains inside the system that establishes the norm — as a trace that cannot be totally eliminated because it is essential to the functioning of the mechanism of exclusion. Thus subjective opinions and habits, the real enemy of an efficient system, constitute, as the human element, the very precondition of a system. Seen from this perspective, the concept of communication implied here is formed in between the conceptual "whole" of a regimented system, on the one hand, and the actual forms of its realization, on the other, and which together determine the field in which it can appear as an object and a means of action. This key point is worth pursuing.

74 Taylor, *Scientific Management*, 119.

75 Thus according to Peter Drucker, for instance, Taylor's is perhaps "the most powerful as well as the most lasting contribution America has made to Western thought since the Federalist Papers" (cited in Kanigel, *One Best Way*, 11).

It was by and through the scheme introduced by Taylor that a line between a system and its "outside" was drawn. For Taylor, the notion of a system was already a pre-given for thinking about work and communication. As he saw it, "in the past the man has been first; in the future the system must be first."[76] The system represented what was rational, ordered, and efficient, whereas its outside comprised all that was not in line with these characteristics, things and processes external to the system that endangered the order so laboriously created and maintained. What exactly was this externality, this "other" side of the system, then, with respect to Taylorism? During the second half of the nineteenth century the culture of the workplace was characterized by the autonomy that craft workers exercised over their labor, and this was what Taylor wanted to change.[77] By the reorganization of work he sought to efface 1) practices based on "habit," "hunch," and "rule of thumb," 2) subjective judgments, individual opinions and initiatives, "guess work," 3) "unnecessary movements," "slow movements," and what he called as "soldiering" (pretending to work), and, finally, 4) "dispute," "discord," and "strife" between the boss and the workmen.[78] It was the task of scientific management to seek to extract, from this disintegrated and inefficient mass of traditional knowledge and physical movement, the exact conceptual structure, and then to reduce it to "laws, rules, and even to mathematical formulae."[79] The collective culture still prevailing among craft workers would be rooted out and replaced by incontestable, absolute norms. This can be perceived as a crucial move in the politics of communicative governance and the governance of communication, being incessantly reiterated, applied, and elaborated later. Through it, the mechanism of a system was synthesized and formalized out of its previously heterogeneous parts and described in terms of its intrinsic nature, seen as a logical whole. For the first time in a "scientific" sense, communication assumed a thoroughly rational structure.

An impersonal, uniform structure was to displace time-consuming and incoherent traditional practices that, from the point of view of a rational system, usually only led to confusion, disagreement, and waste. "Scientific management will mean," argued Taylor, "the elimination of almost all causes for dispute and disagreement between men."[80] These differences of opinion and disordered, incongruent practices constituted, according to Taylor, the ultimate enemy of a perfect system. And it was communication, first and foremost, that provided the means of overcoming the

76 Taylor, *Scientific Management*, 7.

77 Lucy Taksa, "Scientific Management: Technique or Cultural Ideology?," Reprinted in *F. W. Taylor. Critical Evaluations in Business and Management*, vol. I, ed. John Wood and Michael Wood, (London: Routledge, 2002), 389.

78 Taylor, *Scientific Management*, 25, 80, 100, 104, 114, 117, 119, 140, 142; Taylor, *Shop Management*, 30–6; Frederick Taylor, *Testimony before the Special House Committee*, in *Scientific Management: Comprising Shop Management, the Principles of Scientific Management, Testimony before the Special House Committee*, ed. Kenneth Thompson, vol. I of *The Early Sociology of Management and Organizations* (London: Routledge, 2003), 24, 30–1, 41, 45, 97, 109, 116–21, 128, 142. Bernard Doray has listed a number of proscribed types of behavior from this period, including leaving the assigned work position, wandering, loitering, acting disrespectfully, whistling, singing, reading, talking, fighting, and combining (Bernard Doray, *From Taylorism to Fordism: A Rational Madness*, trans. David Macey (London: Free Association Books, 1988), 30–2).

79 Taylor, *Testimony*, 40. See also Doray, *Rational Madness*, 83; Kanigel, *One Best Way*, 217.

80 Taylor, *Scientific Management*, 142.

state of disorder, for "more than all other causes, the close, intimate cooperation, the constant personal contact between the two sides [a manager and a workman], will tend to diminish friction and discontent."[81] The infinite exchange of views to be eliminated, in this connection, is precisely what could be called politics. From the point of view of Taylorism, politics made the functioning of a system inefficient, expensive, or dishonest.[82] Seeing it as a hindrance to working order, Taylor attempted to overcome politics not by another set of politics as other reformers did, but by truth. That is, he aimed at identifying the most effective method for any task by discovering the inherent laws that controlled it.

This necessity to eliminate negotiations by establishing a law forged a new relationship between a system and a nonsystem. Even though a law-based order took the place of disorder, the latter could not be completely eliminated for it was essential to the existence of the former. Upon further scrutiny, the outside of a system was defined as outside concurrently with the implication of the method — that is, when the constituent parts and processes were reconstructed on the basis of their necessary logical structure. Consequently, what happened here, in fact, was that the outside remained part of inside, part of the system, in the sense that the latter could not be considered as completely bereft of human imperfection. Rather, through methodical principles it was possible for the system to regulate inconsistencies in a rational manner. So, what is at stake here is the administration of internal uncertainties by exploiting and refining relevant objects with respect to two things; namely, the fixed goal of a system and what was external to it. In a word, it was a question of constructing a norm in terms of efficiency.

By drawing a demarcation between order and disorder, or between system and nonsystem, method or rule redefines itself time and again through exclusion — that is, negation. The mechanism of exclusion retains order by interpreting it as the general or normal, out from which the special, divergent, or individual is separated and excluded. At the same time the necessary and essential is distinguished from what is incidental and variable as in scientific method, from which it receives its theoretical driving force. Thus the order sets the norm, against which everything is measured.

This said, it is clear, though, that what fueled the appeal and spread of this approach was precisely the vision of the possibility of reconstructing the whole social system as a *machine*.[83] This conception had its roots in late-nineteenth-century America, when the image of the machine was used to explain how diverse processes, activities, and structures were interlinked to form a functioning system.[84] It was Taylor, however, who gave work and communication, understood in terms

81 Ibid., 143.

82 Karl, *Executive Reorganization*, 19.

83 In the words of Taylor, "[t]he mechanism of scientific management is a big engine" (Taylor, *Testimony*, 189).

84 Marcus and Segal, *Technology in America*, 140; Henry Elsner Jr., *The Technocrats: Prophets of Automation* (Syracuse: Syracuse University Press, 1967), 134; Haber, *Efficiency and Uplift*, 119, Wiebe, *Search of Order*, 146; David Hart, *Forged Consensus: Science, Technology, and Economic Policy in the United States, 1921–1953* (Princeton: Princeton University Press, 1998).

of a system, a mechanistic foundation in a "modern" sense. It did not view these terms as separate but in contrast tied the manager, workers, factory, machines, and the actions and routines needed for the work to one and the same continuous structure. It also forged a new relationship to time available. The most economical rhythm of motion became the standard for expenditure of time and removed any discretionary power to take turns working. As a consequence, man and his work, as well as the means and methods of production were all conceived as part of a consistent mechanistic system through physical and functional separation of tasks, standardization of production, and through bringing time in as an essential factor in production. In this way, if technology was previously viewed, to some extent, as a resource external to work, now Taylor built it into the organization of work itself.

The mechanistic approach inherent in Taylorism, and, in the main, also in its successors, constructed a consistent, recursive chain of action that, in the specific circumstances of any given case, would establish the one best, unchanging way to get a task done. Once the chain was established, there would be no need to change it. Only persistent monitoring and communication would be needed to check that the planned line of action was maintained and no deviations or interruptions occurred. It was this rigid methodical approach, remaining the same in every instance, that, if followed strictly, would demonstrate the immanent order of communication, and its connection to the production process.

In this sense Taylor can be considered to have built on the legacy of Ward, for he formulated a method for the systematization of functional action, thereby trying to define the very logic, the hidden mechanism of communication at the core of this action. In a Platonic vein, Taylor sought to identify a simple, rational, and true foundation underlying the heterogeneous and variable world of appearance and to subordinate this apparent multiplicity to a founding law. By deriving this law through a controlled variation of isolated elements, he applied scientific method to the field of functional communication in order to make it more effective. The uncovering of the elementary elements of a given activity and their rearrangement in accordance with the requirements of the tasks in question was seen as a prerequisite for the efficient organization of communication. By undertaking the task of specifying the scientific laws of communication, Taylor can be said to have laid the foundation for thinking about communication as a factual and fully determinable entity at the service of administrative functions. In this respect, he became the central theoretician of communication of industrial-age America.

What was crucial to his concept of an efficient organization was the idea of *repetition*: communication consisted here of regular and controlled reiteration of similar kinds of routines. In this context, every routine performed a certain function within a given productive whole, and communicative operations were then standardized and made interchangeable like machine parts, according to the "American System" in the second half of the nineteenth century. This also necessitated the unification of technical communication terminology concerning the telephone, the telegraph, and the radio.[85] An essential feature of this procedure

85 See Frank Wolff, "Why Standardize?," *Telephone Engineer* 22, (1), (July 1919); "Annual Report of

was that a recurrent, predetermined series of action and communication routines constituted the foundation for efficient production. Because the contents of these acts were already preestablished in relation to the desired outcome, more important than the content of communication was its formal, controlled repetition. Communication assumed here, for the first time, the character of a scientifically produced and controlled function, being subordinate to the needs of the industrial organization it belonged to.

It is possible to perceive this idea of repetition, as put forward by Taylor, as epitomizing the industrial interpretation of communication by reducing the number of its diverse realizations to its lowest common denominator – that is, by reducing a multiplicity to a standard. What Taylor did to work and communication was congruent with what Henry Ford did later to automobiles and transportation in general.[86] The concept of repetition formulated a crucial principle of the modern industrial society. Its effectiveness lay in the fact that it was not restricted to any particular application but in theory was applicable to *any* industrial purpose. Hence it became a generalized tool that not only advanced industrial production but also was itself a product of industrial functions, for it was expressed through them in at least two senses. First, it did not bring into them anything that they did not already contain, but only reorganized their internal sequences of action according to the law immanent to them. Second, and more essentially, it formulated operational laws of movement from the perspective of industrial institutions, and in so doing remained inside the industrial order. What was fundamental, of course, was that this idea of producing and modeling serial processes became possible on the basis that this modern industrial order had laid down in the late nineteenth century. This is how communication pushed ahead along the road of "standardization and efficiency," in the words of telephone historian Herbert Casson.[87]

It is important to note that the efforts to systematize work and communication were not restricted solely to efficiency discourses but developed into a more general framework. Scientific management was consequential in its effects on organization and governance in American society, becoming, according to Stephen Waring, the political philosophy of bureaucratic government.[88] For public administration,

the Telegraphy and Telephony Committee," *Telephone Engineer* 22, (2), (August 1919).

86 David Hounshell pointed out that, despite the commonplace assumption that Fordism relied on Taylorism, there is much reason to doubt this, particularly because unlike the latter, which sought to improve the efficiency of workers, the approach of the former was to replace labor by machinery, typically a moving assembly line. On the relationship between Taylorism and Fordism, see David Hounshell, *From the American System to Mass Production, 1800–1932: The Development of Manufacturing Technology in the United States* (Baltimore: Johns Hopkins University Press, 1984), 249–53. Frank Gilbreth, a prominent figure in the early scientific-management movement, observed that Taylor's ideas on scientific management should not be confused with Henry Ford's concept of assembly-line production. Ford himself also denied the connection (Wren, *Management Thought*, 170, 263–4).

87 Casson, *History of the Telephone*, 169.

88 Stephen Waring, *Taylorism Transformed: Scientific Management Theory since 1945* (Chapel Hill: University of North Carolina Press, 1991), 12. On the consolidation of bureaucratic authority in state

it was an "inspiration, model, and source," in the words of Dwight Waldo,[89] as this new discipline essentially grew from the scientific management movement.[90] The wave of rationalization permeated gradually a whole range of governmental practices with the introduction of management techniques related to reorganization of offices and rationalization of workflow.[91] Moreover, communication was interpreted along the lines articulated by Taylor in industry that made a business of it: the ideas of scientific management were an essential way of comprehending the nature of the industrial production of communication. AT&T, for instance, actively implemented and applied the theories of scientific management.[92] It seemed that if social development was, to a certain extent at least, capable of being planned by using "scientific" methods, so the telephone system, insofar as it were to keep pace with development, must be similarly directed as a single entity.[93] As Roger Burlingame, a historian of technology, put it in the 1940s, AT&T became "a symbol of the collective phase,"[94] for it crystallized the impetus toward scientific administration in a way that created a model for social governance in general. This is also why the telephone as a system became important as an image.

The idea of the telephone company as the model for governance had been presented already by Charles P. Steinmetz, the most famous engineer after Taylor. According to him, the government should take as its model the corporation, and thus introduce supposedly disinterested and unambiguous methods for decision-making.[95] AT&T (or in general "the Bell System") was proposed as an explicit model for the successful management of an immense organization especially in the *Technocracy Study Course*, a book that encapsulated the central ideas of the technocracy movement in the 1920s and 1930s. This movement advocated scientific analysis and the use of technology as methods through which human welfare could be optimized. With Howard Scott as its central figure, it attracted several hundred adherents and formed part of the political momentum that swept

institutions, particularly in military service and public education, by way of Taylorism, see Char Miller, *Taylored Citizenship: State Institutions and Subjectivity* (Westport, CT: Praeger, 2002).

89 Dwight Waldo, *The Administrative State: A Study of the Political Theory of American Public Administration* (2nd edn), (New York: Holmes & Meier, 1984), xxxiii. He goes even so far as to propose it having been "a significant, perhaps a major, input to the sociopolitical realm of the United States" (xxxi). See also Wilson, *National Planning*, 26.

90 Donald Kettl, *The Transformation of Governance: Public Administration for Twenty-First Century America* (Baltimore: Johns Hopkins University Press, 2002), 10. See also Wren, *Management Thought*, 247.

91 Karl, *Executive Reorganization*, 148.

92 Pool, *Forecasting*, 59, 64; see also C. H. Coar, "Encouraging Efficient Employees," *Telephone Engineer* 9, (4), (1913); "Efficiency Engineering in the Telegraph Service," *Telegraph and Telephone Age* (January 1, 1916). For histories of AT&T and the telecommunications field, see, for instance, N. R. Danielian, *A.T.&T. The Story of Industrial Conquest* (New York: Vanguard Press, 1939); Joseph Goulden, *Monopoly* (New York: G. P. Putnam's Sons, 1968); Gerald Brock, *The Telecommunications Industry: The Dynamics of Market Structure* (Cambridge, MA: Harvard University Press, 1981); Cohen, "Telephone Problem"; Mueller, *Universal Service*.

93 Burlingame, *Engines of Democracy*, 118. This view was linked to the attempts to legitimate the concept of "natural monopoly," in the light of which AT&T came to be conceived over decades, see e.g., Vail, "Public Utilities."

94 Burlingame, *Engines of Democracy*, 120.

95 Charles Steinmetz, *America and the New Epoch* (New York: Harper, 1916), 156. See Jordan, *Machine-Age*, 57–8.

Herbert Hoover into office.[96] According to the *Course*, the Bell System established a pattern by way of which rational coordination was practiced successfully against a background of continuous and uninterrupted transformation of the whole system, epitomizing the efficiency of modern industrial organization.[97] This was not, however, a novel perception as the dynamic structure of the system in question was noticed as early as 1908.[98] As business became widely regarded as the model for good government, this giant monopoly continued to attract the notice of contemporary commentators.[99]

The question of social change was an immediate political problem of the time, which is why it was vital, as economist Wallace Donham said in 1932, "to introduce into our American industrialized social organism methods of dealing with change without loss of fundamental economic and consequently social stability."[100] Here, it should be noticed that the question presented itself naturally in terms of lack of a method, which, when found, would give organization of the social a solid foundation. For the technocratic branch of progressivist thinkers, in any case, the Bell System came to serve as a model for a dynamic organization that mastered its own development through formalization of action.[101] In so doing it exemplified the basic design, as anticipated and planned by the technocrats, for the larger social realm to come.

Thus the technocracy movement focused on the methods formulated by scientific management — standardization, the interchangeability of elements, and the maximum economy of a set of actions — which had been introduced into some, although nowhere near all, of the key institutions of modern American society, from the beginning as applicable to society as a whole. Even though Taylor had also thought of his system in relation to society and not just to the industrial establishment, the technocrats' project was explicitly to carry out rational planning directed at society as a whole. For the technocrats, however, there was only one question of importance regarding communication; namely, "of two equally effective modes

96 Cowan, *Social History*, 217. Although the influence of the technocratic movement on American society remained limited, many familiar technocratic themes resurfaced in the 1950s and 1960s (see Elsner, *Technocrats*, 216–17). For Howard Scott's ideas, see Howard Scott, "Technology Smashes the Price System," *Harper's Magazine*, 166, (January 1933); Howard Scott et al., *Introduction to Technocracy* (London: John Lane, The Bodley Head, 1933); Howard Scott, "America Prepares for a Turn in the Road," *Technocracy* 1, (7), (1935).

97 The Bell System, according to the *Course*, "maintains in continuous operation what is probably the most complex, interconnected array of physical apparatus in existence," by "continually changing the apparatus with which it has to deal, and remoulding the organization accordingly" (*Technocracy Study Course* (Technocracy Inc., 1934), 221–2). See also Elsner, *Technocrats*, 127–8; Akin, *Technocracy*, 86.

98 John Mumford, "This Land of Opportunity: The Nerve-Centre of Modern Business," *Harper's Weekly*, (August 1, 1908): 22.

99 Casson, *History of the Telephone*, 140–1; Sydney Brooks, "Politics of American Business," *North American Review* 93, (May 1911). In 1939, N. R. Danielian considered the Bell System as "the prototype [. . .] of organized business" (Danielian, *Industrial Conquest*, "Introduction"). See also Hanson, *Democratic Imagination*, 242–3.

100 See Wallace Donham, *Business Looks at the Unforeseen* (New York: Whittlesey House, McGraw-Hill Book Co., 1932), 26.

101 It was during this period that the business corporation was taken in the writings of the reformers as a generic model for governance in various organizations, including the national government: it thus became the framework for administration (Haber, *Efficiency and Uplift*, 107; Lustig, *Corporate Liberalism*, 97).

of communication which has the least energy cost per unit?" The unit, in turn, was defined as "a given number of words transmitted a given distance."[102] Here, the idea of communication as purely the transmission of preexistent, independent contents was given its strongest and most resolute expression within the heritage that can be traced back to Ward. Communication became exclusively a matter of organizational efficacy, for it was of interest to the technocrats only in terms of its effects on the replacement of manpower by advancing the automatization and mechanization of society as a whole. Clearly, in the hands of these thinkers it was rendered a completely self-enclosed instrumental system: it was not only viewed as being subordinated to the industrial aims it seemed to serve and express but also as receiving its positivity, value, and social meaningfulness through the predetermined and unquestioned objectives of the system itself. So, its seemingly infinitely variable forms could in principle be identified and determined according to their social function and the level of energy consumption, and thus produce the best communication model for each purpose required.

The technocrats, who largely received their original inspiration from Thorstein Veblen's book *The Engineers and the Price System*, published in 1919, held that because contemporary social problems, as they saw them, were predominantly technological in nature, it was the engineer and not the politician who must take over the management of the country.[103] This opinion reflected the general sentiment among progressives that the growing complexity of the industrial society made the expert a prominent figure, although only the technocratically oriented branch was willing to grant the expert authoritative powers.[104] By articulating the engineer's point of view, the technocrats capitalized on the prevailing impulse to use clear, objective methods to reach unambiguous solutions and to eliminate political dispute. Politics, as they saw it — or any social or cultural factor for that matter — could make no contribution to resolving questions of fact, which should therefore be left to technical experts.[105] "The stoking of a bunsen [*sic*] burner, the

102 *Technocracy Study Course*, 1934, 255. See also Scott, "Price System," 130–2, 142.

103 See Allen Raymond, *What Is Technocracy?* (New York: Whittlesey House, McGraw-Hill Book Co., 1933), 174. Of course, many social theorists and sociologists from the early twentieth century, such as Charles Cooley, seeking a disciplinary model for the science of society disregarded Veblen's ideas and turned to the capitalist market as a central progressive force. However, this did not shatter the sociologist belief in social control, based on scientific knowledge, as the other major factor in what they saw as modern progress (Ross, *Origins*, 149, 223–52). For corporate-liberal thought, see also Martin Sklar, *The Corporate Reconstruction of American Capitalism, 1890–1916: The Market, the Law, and Politics* (Cambridge: Cambridge University Press, 1988).

104 This branch included, at least in some phases of their careers, figures such as Louis D. Brandeis, Herbert Croly, Walter Lippman, and Woodrow Wilson (see Haber, *Efficiency and Uplift*, 75–98; Lustig, *Corporate Liberalism*, 152–3, 202–3; Edward Stettner, *Shaping Modern Liberalism: Herbert Croly and Progressive Thought* (Lawrence: University Press of Kansas, 1993), 60, 62; Pursell, *Machine in America*, 214, 217).

105 Elsner, *Technocrats*, 30; Akin, *Technocracy*, 1–5. By stressing the engineering method as a model that could replace social contention, rational reform movement wanted to eliminate, among other things, the power of tradition, irrational beliefs, dumb luck, the existing relations of opposition and oppression, partisanship in appointments and "politics" in the practice of government, imprecise qualitative judgments, partisan infighting, favors and debts, empty rhetoric, the profit system, emotions, and a "wealth of discussion" (Jordan, *Machine-Age*, 39, 40, 85, 114, 115, 120, 128).

stoking of a boiler, the stoking of the people of a nation, are all one problem," declared the *Course*.[106] Whereas politics was subordinated to free market in the classical liberalism, in the context of a new, industrial liberalism it had a tenacious disposition to become subjected to the "operating requirements" of technology.

It is clear that community for these thinkers was not a *political* community, for it did not allow disagreement or debate — that is, questioning of the conditions of commonality, the precondition of politics. They held that any competent functional organization "*has no political precedents. It is neither democratic, autocratic, nor dictatorial.* It is determined by the requirements of the job that has to be done."[107] What the technocrats wanted to do was to erect a machinery that would recognize only one norm — that of engineering rationality and efficiency. As the disinterested and impersonal methodological approach appeared the only valid basis for this, the uncontroversiality of the method brought about the apoliticality of governance. The engineering worldview, although pushed forward by technocrats, was by no means limited to the technocracy movement. Thurman Arnold, for instance, best known for his trust-busting campaign as assistant attorney general, reflected this common outlook in his widely read book *The Folklore of Capitalism*, written in 1937, in which he deplored the fact that when dealing with social problems, as opposed to physical epidemics, people were still fettered with moral preconceptions.[108] Because social matters largely became regarded in this discourse as questions related to the choice of appropriate rules of conduct, politics became an interlinked set of practices and precepts, a body of instructions rather than an argumentation. Here, instructions are to be understood not only as a set of precepts concerning the functioning of the social apparatus but also as the product of the "machine-likeness" of the society itself, as a response to questions that became expressible only within this conceptual context.

What seemed to be the logical result of the development of science and technology was an efficient yet nondemocratic government. In accordance with the principles of scientific management, the technocrats' goal was to replace the conflicts of competing interest groups and the "irrationality" of democratic politics with mechanical rationality, functional standards of judgment, and preplanned administrative systematics. Thus the efficiency schools rejected "both politics and the political," according to administrative historian Barry Karl, and centered on the rational and economic system.[109] It is not that politics would be completely removed or eliminated, though. It is rather that rational technology would engender a regime of truth wherein politics would be exhausted through

106 *Technocracy Study Course*, (1934), 116.

107 *Technocracy Study Course*, (1934), 224; emphasis in original.

108 Thurman Arnold, *The Folklore of Capitalism* (Washington, DC: Beard Books, 2000 [1937]), 86. Of course, the inevitability that made all problems technical was nourished by a certain political *ethos* — something the rational reform movement had intended to eliminate in the first place; see Louis Hartz, *The Liberal Tradition in America: An Interpretation of American Political Thought since the Revolution* (New York: Harcourt, Brace & World, 1955), 270–1.

109 Karl, *Executive Reorganization*, 24, 150. See also Frank Fischer, *Technocracy and the Politics of Expertise* (London: Sage, 1990), 22–4; Lustig, *Corporate Liberalism*, 171–2; Pursell, *Machine in America*, 204.

unambiguous methods and procedures. Politics, insofar as it existed in the plans of these technocracy-minded thinkers, would certainly not be defined in terms of contention or dispute, but rather in those of efficiency and stability.

Rationalizing Communication

The rational reform movement initiated the project that placed communication and the communicative order on a scientific footing, wherein applied sciences were combined with the engineering model. This movement was concerned with institutionalizing communication in a way that would correlate with industrial organization, and in this sense also with modification of the conditions of communication in line with this objective. These conditions appeared no longer as a set of unchanging natural laws but instead for the first time as a modifiable order, subordinate to rational design. But as communication had assumed an unprecedented positivity and social effectiveness resulting in the systematic study of its mechanisms and channels, so the conditions of communication were seen in this instrumental light, which placed its ontological dimension in a technological context. In this way, as the progressives articulated society as an administrable whole, the need to take the reins pushed the dimension of communication as an instrumental means to the foreground. This field of growing technicization, of course, laid the very foundation for the general possibility of the idea of being able to control the direction of society.

Communication research took shape in and through this perspective of administration and control; in fact, it was, in the final instance, this very perspective that made it possible as an institutionalized practice. As political author Herbert Croly expressed the view that the extreme concentration of industrial and financial power had resulted in the disintegration of political order and that what should be done was a restoration of social stability and a sense of national unity,[110] communication as a social process was immediately politicized. So, communication research was given a social function within the political movement that was only strengthened during World War II: to provide knowledge about communicative means, practices, and effects in order to maintain the stability of the political system.[111] Here, there was a notable break with communication as understood within the framework of the natural-rights doctrine of liberty of contract, which positioned it as an interindividual mechanism. Although also aiming at engendering a sense of unity, this mechanism proved unfeasible as it apparently led to the dissolution of political order. It conflicted precisely with what the progressives were struggling to achieve: it was what they wanted to abolish in order to give way to a concept that was predicated on collective, not interindividual processes.

On the other hand, the demands of industry resulted in the development of methods and techniques that would aid in rationalizing production systems. Communication, particularly new communication technologies, if they seemed to

110 Scott Bowman, *The Modern Corporation and American Political Thought: Law, Power, and Ideology* (University Park: Pennsylvania State University Press, 1996), 81.

111 Hardt, *Critical Communication*, 17.

support this objective, was integrated sooner or later into this scheme. This is also why Taylorism must be seen as applying directly to communication, too. It is also important to notice that it was not limited to purely industrial activities, but spread throughout society and grew as a common conceptual horizon within which the question of rationalization was expressed and executed. Given this, it created an environment also for prosecuting broader political interests along the lines of progressive thinking. As Scott Bowman has pointed out, underlying the practical application of the idea of social balance was the norm of efficiency. This was the source of the attractiveness and viability of Taylorism and other efficiency schools for political thought; they supported the political cause by providing mechanisms to restore social balance, for free contractual relations proved unable to maintain it.[112] Stability could be sustained only by efficiency; where previously there was a loose set of connections between organizations and practices, one should create a comprehensive and consistent social machinery. This is how social stability was combined with a hierarchical principle of organization,[113] and with this conceptual bond communication as a social process was connected to a hierarchical system. Although this kind of total system was not, of course, brought into being in any comprehensive social form, Taylorism provided a model that framed the political discussion on how to retain the beneficial consequences of efficiency without its detrimental effects.

Thus, in his book *Progressive Democracy*, which sought to provide a program for the progressive movement as a whole, Croly argued for the creation of "industrial democracy" within a system of "scientific management."[114] These two terms were not independent of each other but, on the contrary, seemed to be inherently linked by their common goals of efficiency and stability.[115] If scientific management was first and foremost a political technique, then its practical application was not restricted to management of business but was felt essential for industrial democracy in general. There was also a second reason why scientific management seemed to be suitable for Croly as a political technique for maintaining and engendering democracy. This is that scientific management — not unlike Croly's progressivism — rested on the idea of a single whole. For many progressives, Croly included, harmonious social structure implied a concept of society as an interrelated and efficient totality.[116]

Now, along with the move from an individual to a group, the telephone — together with other communication systems — appeared as a technique that would engender communality. If communication constructed the community, as was acknowledged, then by controlling the means and structure of communication one could keep the social as one governable whole. This idea, although not often explicitly articulated, was reflected in various forms in contemporary developments.

112 Bowman, *Modern Corporation*, 84.
113 Bowman, *Modern Corporation*, 89; Hanson, *Democratic Imagination*, 259.
114 Herbert Croly, *Progressive Democracy* (New York: Macmillan, 1915), 378–405. See also Stettner, *Modern Liberalism*, 92.
115 Bowman, *Modern Corporation*, 87–8. See also Hanson, *Democratic Imagination*, 253–5.
116 Lustig, *Corporate Liberalism*, 136; Bowman, *Modern Corporation*, 89; Eisenach, *Lost Promise*, 131.

The most prominent of these were clearly the drafting of the Interstate Commerce Act in 1887 and the Sherman Antitrust Act in 1890, through which freedom of contract was subordinated to national regulation, as it seemed out of touch with reality in the eyes of the progressives.[117] One of the most important regulatory statutes in the twentieth century was the Transportation Act of 1920, the main purpose of which was to prepare a plan for an efficient network of roads to increase competition and to foster consolidation of the nation's telephones into a single system. Crucial to this was the idea, as Alan Stone noted, that the Interstate Commerce Commission would employ scientific-management techniques to produce a highly efficient system.[118] The same procedure — of maintaining social stability through increasing regulation — was proposed also in connection with social conversations over telephone lines.

It found expression in early discourses on the telephone, especially those produced by telephone companies and that first attacked social conversations over telephone lines. Chat, so telephone companies and probably most of the people at that time believed, only kept lines busy and conflicted with the idea of their economical use. Therefore, in order to maintain the "efficiency" of the communication system, chat had to be abolished. Thus, the editorial of the *Telephone Engineer* in April of 1920, entitled "Making the Public Efficient," for instance, regarded proposals to "eliminate such abuse of the telephone privilege" as a "brilliant reform [idea]," though judged it as probably "unworkable."[119] Similarly, J. S. McCulloch, the vice president of the New York Telephone Company, was reported as having proposed that "superfluous chatter over the telephone be eliminated."[120] Although it is obvious that this urge was motivated partly because of the tendency to continue to think of the telephone in terms of its telegraphic heritage as strictly a tool of business,[121] another reason for this was certainly the desire to set some guidelines for the communicating community. One could regard this episode as

117 Garraty, *New Commonwealth*, 118–22; White, *Republican Era*, 5; Morton Keller, "The Pluralistic State: American Economic Regulation in Comparative Perspective, 1900–1930," in *Regulation in Perspective*, ed. Thomas McCraw (Cambridge, MA: Harvard University Press, 1981), 66–71. However, these Acts were denounced by jurists advocating laissez-faire liberalism as being in conflict with the assumed natural laws of trade and the principles of the political economy (see Fine, *Laissez Faire*, 139).

118 Stone, *Public Service*, 271. See also Skowronek, *New American State*, 283; Morton Keller, *Regulating a New Economy: Public Policy and Economic Change in America, 1900–1933* (Cambridge, MA: Harvard University Press, 1990), 79.

119 The editorial continued by lamenting that "[e]xasperating and discouraging as it may be to know that the wires are tied up by useless gossip while valued business subscribers clamor for connections, technically the gossipers have as many and as valid rights as the business men. That we are all sorry this is so does not make it any the less true" ("Making the Public Efficient," *Telephone Engineer* 23, (4), (April 1920): 11).

120 "Telephone Chief Wants Gossip Prohibited," *Telephone Engineer*, 23, (4), (April 1920): 14. The discussion continued in the following year, when the editorial of the same journal stated that "[t]he telephone was not invented and perfected to be used for social gossip. The real purpose of this modern convenience is to annihilate distance, to place people in speedy and direct communications with each other and to serve as an inestimable asset in the world of business especially. [. . .] If you must tell a long story over the telephone make it a serial and run it in episodes throughout the day" ("Using the Telephone," *Telephone Engineer* 24, (4), (April 1921)).

121 Sidney Aronson, "Bell's Electrical Toy: What's the Use? The Sociology of Early Telephone Usage," in *The Social Impact of the Telephone*, ed. Ithiel de Sola Pool (Cambridge: MIT Press, 1977), 27; Irwin Lebow, *Information Highways and Byways: From the Telegraph to the 21st Century* (New York: IEEE Press, 1995), 56.

indicative of the need to maintain the governability of communication — at least by establishing limits — following the ideals of rationalization. Commentators on both sides of the Atlantic welcomed the telephone's influence on human intercourse, as not only was speech "standardized and economized" because of it but also it taught people "a desirable frankness and simplicity."[122] The underlying rationale was aptly encapsulated by Herbert Webb, writing in 1910: "The essence of the telephone service is efficiency — and here efficiency means rapidity combined with accuracy and reliability."[123]

The rise in efficiency discourses over this period resulted in chat appearing as the enemy of well-ordered communication facilities: like politics, it represented the same endless exchange of opinions that only seemed to undermine the most economical use of the system. This was, of course, backed up by the needs of business, as conversation over telephone lines for any purpose other than business was regarded as a waste of time and money.[124] The influence of Taylorism is thus clearly visible.

This implied not only administration of communication but also administration through communication. Communication took central stage in the organization of community, and the technical means of doing this were naturally in the spotlight. As already mentioned, the primary responsibility for constructing the communicative domain was given to communications companies, which thus formed the vanguard in seeking to tie the community with communicative bounds into a single consistent whole. Of course, engendering forms of the social must not be understood narrowly, here, as in the United States this meant largely strengthening and unifying the national basis of communication and, as is the case here, supervising and preventing unwanted practices from taking place. As this activity, however, took rather prolonged forms that did not change until after World War I when the use of the telephone for social purposes began to be actively encouraged, systematic monitoring of the use of communicative means must be viewed in its politicality: it expressed a concern of the social viewed as a whole. Yet communication did not appear as a major political issue as such, particularly due to the inclination to see it through intrinsically rational rather than political systems.

Without doubt, Taylorism set the framework for "scientific" problematization of management and communication (and certainly on communication as management as well as management of and through communication) for years to come. For communication, at least within the tradition outlined here, this implied that it became difficult to separate communication from managerial problems: the

122 John Lee, "The Telephone and Social Reconstruction," *Telephone Engineer* 18, (2), (August 1917): 65.

123 Herbert Webb, "The Development of the Telephone in Europe," in *The Telephone Systems of the Continent of Europe*, ed. A. R. Bennett (New York: Arno Press, 1974), 49.

124 Michèle Martin, *"Hello, Central?" Gender, Technology, and Culture in the Formation of Telephone Systems* (Montreal: McGill-Queen's University Press, 1991), 115. The telephone was not advertised as a means to socialize until the 1920s; see Fischer, *America Calling*, 75–80; David Nye, "Shaping Communication Networks: Telegraph, Telephone, Computer," *Social Research* 64, (3), (1997): 1071.

modern history of communication was closely interwoven with the politics of governance. What is more, problems relating to management and control were precisely what allowed communication to appear as a distinct concept and a category of action; it was through discursive chains of management and control that communication was conceptualized — first as a tie among individuals, then as a mechanism of the social itself. In both cases, though, communication served as a method of engendering and organizing the social as a whole and, at the same time, assumed a status of its own in the course of this very process.

Of course, the perspective within which communication was perceived became wider as well. The telegraph incorporated communication for the first time into the sphere of market economy by rendering telecommunication a saleable product. This highlighted not only the status of communication as a new industry but also its growing role in the commercial and financial circles of society. During the age of the telephone, for its part, thought about communication was broadened from the perspective that restricted it to the special connection communication had with business. Whereas previously telecommunication had been largely thought of in terms of private enterprise, now it began to be thought of more and more as a general social category that enabled consistency in a society. This was, of course, part of the transformation in which the concept that society consists of unified and autonomous subjects and of their interrelations (regarded predominantly as economic), and for which communication was merely instrumental, was replaced by a concept that communication was constitutive of the whole society and of the intersubjective relations characteristic of it.[125] Yet after Taylor, thinking on communication was closely connected to the way contemporary capitalism operated. Communication, it is true, would not have been possible in its modern forms without capitalist markets that, generally speaking, defined the domain within which communication became an object and means of systematic thinking. By starting from corporate practices and needs and not from communication as such, Taylor gave this relationship a theoretical basis that permanently intertwined the history of communication with that of capitalism. Although impossible to disentangle from communication, the domain of corporate capitalism, here, determined the terms in which communication was conceptualized.

Scientific management was not a philosophy of capitalism but rather one that operated within it, "reconstituting the world as the mechanical embodiment of the corporate director's own categories."[126] While reorganizing the chains of productive action it remained firmly inside the horizon of capitalism. What this implied, among other things, was that communication was to be thought of as fundamentally apolitical and, at the same time, as a mechanism to ensure and maintain this apoliticality. It was based on the aim of doing away with disputes over values, relying on differentiation and specialization, with this differentiation and special-

125 The aforementioned report from 1933 expressed this notion thus: "The existence and continuity of social life depend on communication" (Malcolm Willey and Stuart Rice, *Communication Agencies and Social Life* (New York: McGraw-Hill Book Co., 1933), 210).

126 Lustig, *Corporate Liberalism*, 190.

ization coming to determine communication as both technique and community. Yet it also created problems without being able to solve them. Much of the later history of control of and by communication can be seen as a series of attempts to overcome these problems, linked with the tendency in Taylorism to create conflicts and operations that could not be controlled through the means provided by it. While one school responded to the problems by developing participative methods of management in the progressive spirit, another school formulated mathematical and mechanical techniques in order to make bureaucratic management more efficient. In the succeeding chapters, I will follow largely the latter line, as the former school, in general, accepted centralized power, hierarchical controls, and specialized tasks too, the three main tools of bureaucratic technique that unified the subsequent history of communication until the 1970s.[127]

Taylor formulated a way of talking about communication under growing corporate power and the increasing influence of formal techniques. Communication was something to be regulated by rational means, constituting at the same time the mechanism for this regulation. It is important to notice this "at the same time," because the mechanisms introduced emerged, ultimately, out of communication itself. Thus there was no conflict between the objects and means of control, no violence done to the intrinsic nature of communication, as it were, as the way it was organized depended on the elementary constitution of communication, on the inevitable parts and processes of it. Contemporary capitalism was, for Taylor, the context that brought out this communicative structure that varied according to the task in question. It constituted the horizon that not only laid bare but, in fact, produced the practices and series of action Taylor sought to systematize.

It is perhaps possible to consider the cybernetics of Norbert Wiener a kind of culmination of this development, as it is here that communication as a law or a method was given probably its most resolute expression and in which it at the same time was linked with a new, computer-controlled horizon. Let me first, however, elaborate in the next part of this chapter on the component position of the telephone system with respect to the growth of hierarchically controlled organizations. For they were the reference point on which the studies of Taylor and the technocrats — as part of the general thinking that took as its subject the changing conditions of social operations — were focused and which thus constituted the environment within which communication became to be thought about. At the same time they provided the context in and with regard to which it was only natural for Wiener to formulate his theoretical views, which can be seen as a modernization of the model built on and around the idea of a system to correspond to the circumstances and conceptual possibilities of a period of automated technology.

Machines of Communication

It is possible to conceive of the discontinuity caused by and in telephonic communicativity from two distinct perspectives. Like the telegraph, the telephone

127 Waring, *Taylorism Transformed*, 7–8, 17–19; Wren, *Management Thought*, 297.

could be seen primarily as a means for the horizontal distribution and transmission of information, even though this aspect of communication often becomes meaningful only in a context that emphasizes its inherent unifying tendency. Hence it was customary, especially in the progressive tradition, to view horizontal communication as a force for rendering society into a truly moral public. Within this tradition, science and knowledge attempted to overcome social factionalism as well as conflicts and disputes, and it was the task of communication and education to accomplish this general social harmonization. Yet, on the other hand, the telephone — both the technology and the model it provided — can be regarded as closely connected to hierarchical control.[128] In this view, communication imposed law, as opposed to education, as the method of politics.

As part of the same process, the nervous system and other related metaphoric images — such as the telephone exchange as the brains of the system and the telephone as the nerve ending of society — were consolidated in social use, following the models provided by its predecessor in telecommunications. To be sure, society did not become a huge system with its diverse functions and processes linked together and controlled from a single executive center. However, the idea of this kind of organization certainly evoked enthusiasm, mobilizing plenty of energy, and resulted in some very concrete arrangements, practices, and effects. Not only had the idea of a mechanically controlled whole had real consequences for various schemes and procedures but also, and more important, it provided, to a great extent a conceptual perspective in which these schemes could be articulated in the first place. In this way it forged the limits within which the social was understood.

Even though both perspectives on the telephone appeared here as distinct from each other, they should not be thought of as logically or temporally separate but rather as views that are aggregated and arranged, according to their own principles, around an event that is being investigated at any given time. I can define the first case, in which the telephone is seen as a means of horizontal communication, in terms of a relationship, the function and parties of which are not given and the formal occurrence of which is not determined. Instead, in the second case, in which the telephone is viewed as an integral part of centralized structures, its function and parties are determined by the system. A communicative relationship here is a relationship internal to the system, and its functional nature is thus prestructured as regards its formal characteristics.

In both cases communication is conceived in terms of being a more or less coherent and closed whole. There was a change, though, in the way the whole was viewed as we move from the telegraph to the telephone and related discursive forms. Whereas atomistic individuals defined the earlier viewpoint, and communication was seen as an addition between already established agents, now the viewpoint moved from individuals to the broader social group they belonged to,

128 The pattern of telephone operation and regulation was predicated on a hierarchical structure — the preferred solution for most large technical systems in the late nineteenth century; see Andrew Davies, *Telecommunications and Politics: The Decentralised Alternative* (London: Pinter Publishers, 1994), 32; Russell Neuman, Lee McKnight and Richard Solomon, *The Gordian Knot: Political Gridlock on the Information Highway* (Cambridge: MIT Press, 1997), 60.

and communication was more and more regarded as a factor that maintained the integrity of the group or system. It was the social system, that now constituted the basis of thinking about communication, and although communication was a precondition of this system, it became commonly regarded as a means to sustain the unity of this social formation. It is true that telecommunication was a vehicle of social integration during the age of the telegraph too. Yet only with the advent of the telephone, so it seems, the perspective of unification became a scientific basis for thinking about communication within a formal structure.

In light of what we have shown earlier, it can be said that the idea of governance from a distance established itself in its modern sense concurrently with the development of communication arrangements. Once its principle was introduced as part of municipal telegraph systems, the system based on a coordinating center connected with information gathering and executive organs within a hierarchically arranged whole was applied in various ways to numerous contexts, from police and fire-alarm systems to schemes on the national level. It can be argued, however, that this system brought in by the telegraph was a rather superficial phenomenon after all, because the essential communication enabled by this technology remained, in the end, somewhat limited. This changed with the introduction of the telephone. Now telecommunication had the capacity to articulate the industrial structure of society and give it a conceptual form hand in hand with the development in which big industrial conglomerates were raised into cultural icons. At the same time, the emphasis began to move from control based on a dominating center to take into account the role of systematization — in other words, predetermined, formalized relationships within a well-arranged whole. It must be noted here that it is much less a question of the spread of an imaginary uniform idea as such than of the principles of rational planning and scientific method, which gave different expressions to this idea, pervading the whole of society. This is why this model provided by the telephone was so important for thought about society: it became a mechanism — not the only one, of course, but indisputably a very powerful one in any case — for communicative relationships understood as one structure.

This is why the hierarchical aspect is the one that has been emphasized here. Since the beginning of the century, and alongside projects of rationalization and optimization, new forms of communication, with the telephone being without doubt the most important one, have been thought about within the bounds of a rational system. But it is significant that the telephone, as a system, played an integral part in bringing about this rationality in society. This is thus also a junction in which the link between telecommunication and the history of determination and quantification of social phenomena can be seen from another angle. The model based on hierarchical control and on the determination of communicative connections as manifested in and articulated through communications and other technical systems, replaced subjective decision-making with objective rule. The system that entailed this sort of control put a fixed principle in the place of changeable practices based on habit or experience, a principle that also provided the basis for the determination of social phenomena insofar as these were objects of scientific examination. In this sense it was a kind of institutional equivalent to scientific method in the social realm, its application or embodiment: whereas scientific method distinguished what was essential and constant from what was variable and incidental and thus constructed a relationship with truth, the model based on centralization and standardization erected a field of truth in the midst of incongruent and dissimilar practices. Therefore the history of the telephone is inherently tied to the development of hierarchically organized and rationally controlled systems, although its history certainly cannot be reduced to just this development.

Machines of Speech

Scientific management standardized basic units and courses of work and reorganized a given functional whole according to the most economic sequence of actions. In doing this, it anticipated and recorded all possible variations and compressed them into a set of necessary rules. This led to a hierarchical system wherein communicative action was to be controlled by a predetermined method. It was here that a "rational" center, replacing the "subjective" manager in a conventional organization, emerged as a founding *modus operandi*. In Taylor's theory, all managerial power was concentrated on this single locus, the superintendent's office, from which executive work was planned, put into effect, and supervised. In a manner similar to the way the brain directs the movements of the hand, the thinking organ of the organization, it was assumed, told the workers the exact way to do their job. "All possible brain work," Taylor wrote, "should be removed from the shop and concentrated in the planning or laying out department."[129]

In another sense, however, scientific management broke away from the confines of a single dominating authority, as the aim was precisely to decentralize control by placing it in the hands of lower-level managers.[130] Thus, from this perspective Taylorism could be said to have dismantled the very management tradition that concentrates all control into one locus or person, which is characteristic of an economy based on household production: the system is divided into different levels and each level is allocated a distinct functional goal, after which, finally, all the functioning units are tied into a close hierarchical system of responsibilities that brought them into the whole. Furthermore, as I have attempted to show, communication was built into the very structure of this whole, forming a central mechanism by means of which the work was directed and adjusted. It thus became indistinguishable from the actual carrying out of the required tasks, and it was in terms of communication, too, that the decentralized controls operated. The cornerstone of this scheme, as noted earlier, was a rational, repeatable program that identified what was essential from the apparent diversity and multiplicity and defined a system of communication for the instructions to be transmitted. Consequently, instead of creating a single center of management, this scheme

129 Cited in H. Braverman, *Labor and Monopoly Capital: The Degradation of the Work in the Twentieth Century* (New York: Monthly Review Press, 1974), 113.

130 Wren, *Management Thought*, 131.

actually highlighted the impersonal logical structure of any form of work, although the necessary coordinating authority undoubtedly loomed large.

Now, from the background to the history of the idea of a hierarchical system, a whole series of diverse, irreducible, and discontinuous historical constructions can be found, especially as implemented in different military and political systems since the time of Frederick the Great of Prussia. Yet, it was the invention and proliferation of machines, coupled with scientific research, that laid the ground for a hierarchy as a powerful mechanism for distinguishing between "objective" and "customary" ways of organizational performance in modern bureaucracies.[131] The hierarchy manifested the scientific method's principle, which Taylor, for his part, exploited and applied, of substituting subjective decisions by an objective rule. Hierarchical architecture put formal criteria in the place of changeable practices based on custom and experience, and in this respect can be considered as a kind of counterpart, application, or embodiment of the scientific method in the social domain, as already mentioned. Yet, it can be argued, the hierarchy appropriated this method not for truth but rather for efficiency: whereas science separated what was incidental and random from the essential and unvarying, and thus forged a relationship to the truth, as it were, the hierarchy created a realm of efficiency in the midst of diverse and incompatible social practices. Insofar as efficiency sought to shed light on the underlying "laws" of the given process, however, it cannot be separated from ontological questions in any straightforward manner. In both, the necessary and general was derived from what was contingent and arbitrary by the constraints imposed by a rational method. If scientific management apparently dealt only with efficiency, then it did so with a view to truth.

On the other hand, and from a different perspective, insofar as the hierarchy proved to be an influential social-ordering principle as well as a useful conceptual tool in thought about modern societies in general, as Max Weber demonstrated in his contribution on the proliferation of bureaucratic forms of organization, it was to a great extent the network structure of the railroads and the telegraph that prepared the ground for this organizational model as understood in the sense defined earlier. This is because the chaining together of series of actions with enabling physical structures into a consistent operational whole made it possible to arrange interrelationships among events in an integrated way for the purposes of modern society. Thus the ideas of communication as the foundation of governance took shape in, and drew from, the development of early transportation and communication systems, as in the municipal telegraph systems, the network nature of which enabled uniform, system-wide operational control with separate hierarchical levels of regulation. This is undoubtedly the context, too, that must be seen as the point of departure for Taylor as he generalized from prevailing institutional practices and arrangements using science as his guiding canon, the idea of

131 Max Weber encapsulated this by stating that "the fully developed bureaucratic mechanism compares with other organizations exactly as does the machine with the non-mechanical modes of production" (Max Weber, *From Max Weber: Essays in Sociology*, ed. Hans Gerth and C. Wright Mills (London: Routledge, 1948), 214). For an introduction to the concept of bureaucracy, see David Beetham, *Bureucracy* (2nd edn), (Buckingham: Open University Press, 1996).

systematized control with predetermined interconnections into a general managerial principle. Therefore it is possible to see Taylorism as a powerful combination of scientific method and a systematic management. Through the influence of his theory, coupled with the growing dominance of big corporations, the internal functioning of which the former aimed to enhance, Taylor contributed to communication's entrenchment as the organizing mechanism of the industrial order predicated on the concept of vertical lines of communication. Because improved communication systems — especially the telephone system that allowed instant, two-way communication — figured as the key factors in the transformation of the whole communication space in the period in question, they must also be viewed as the forces through which hierarchy as a social category should be analyzed. It was precisely the telephone, according to a historian of this communication system, from which "a new organization of society" evolved.[132]

The telephone played a part in more and more accurately determined needs of governance in modern society by permeating the communal sphere and becoming a mechanism — as part of the increasing professionalization of social organizations — by means of which this sphere could be controlled. Namely, it enabled systematic and institutionalized gathering and distribution of information and quick reaction to phenomena taking place within the range of its operations, in this way fostering the idea of governance from a distance. Thus it became increasingly a part of daily operations in the private sector as well as in public administration, permitting growth in the size of organizations and in the range of processes that could be controlled at a distance.[133] The telephone was soon adopted by the US government, which in 1910 built what was arguably among "the greatest private telephone exchanges in the world."[134] The Weather Bureau introduced it swiftly into its operations as well as many other public bureaucracies like the coast guard, fire department, and the police.[135] In this way, the telephone divided society into distinct, although interlinked and partly overlapping, sectors, each of which administered its own areas of responsibility on the basis of the same principle, familiar from the telegraph, of two-way exchange of information — the gathering of information from the field and the transmission of instructions from the center within the framework of a functional, hierarchically arranged whole.

132 Burlingame, *Engines of Democracy*, 97.

133 Pool, *Forecasting*, 59. For the development of office equipment and bureaucratic techniques during the industrial age, see JoAnne Yates, "Business Use of Information and Technology during the Industrial Age," in *A Nation Transformed by Information: How Information Has Shaped the United States from Colonial Times to the Present*, ed. Alfred Chandler Jr. and James Cortada (New York: Oxford University Press, 2000).

134 Waldon Fawcett, "How Uncle Sam Uses the Telephone," *Telephony* 19, (4), (January 1910): 90. In 1911 the US government telephone network, consisting of a "system of inter-connected private branch exchanges," was said to be "the most extraordinary and most complete system of its kind in the world" ("Government Stations at Washington," *Telephone Engineer* 5, (6), (June 1911): 319). Frank Childs, however, later put the hype in context: "Words at the best were barely intelligible, and as the wire was strung on the same poles as were the telegraph wires, there was much induction, which seriously interfered with the clearness of the message" (Frank Childs, "When the Telephone was Young in Washington, D.C.," *Telephony* 104, (13), (April 1, 1933): 17).

135 "The Telephone in Government Service," *Telephony* (June 1905); Dilts, *Changing World*, 78–82.

In police communication systems in 1910, this two-way communication meant that "the patrolman reports to the desk sergeant from the telephone box on his beat, and then receives any special directions which the chief wishes to give."[136] This principle, with the operational advantages of efficiency and coordination it enabled, also provided a setting for the construction of huge national systems in an age of big conglomerates. Though centric systems were introduced on a large scale in the era of the telegraph, it was not until the advent of the telephone with its superior properties that the foundation for truly professionally organized, nationwide systems was laid. The telephone systems of the government and the War Department were among the largest private exchanges in the world,[137] and it was easy to perceive them — being contemporaries of the consolidation movement that created monopolies and trusts — as having a centralizing effect on governance and communication.[138]

However, as with commercial organizations, they supported simultaneously both centralizing and decentralizing processes in the functioning of these public institutions: if the telephone system made it possible to centralize power at a single point, it also allowed — even necessitated — decentralization and differentiation of operations. Thus the establishment of extensive communication systems as founding institutions of social activity gave rise to, in the very process, necessarily decentralized action as well. The process of centralization can therefore never monopolize the sphere of action of which it is a part. This is not all, however, for the center cannot cease to create differentiations and partitions: every move toward centralization requires new determinations and entails novel differences. In this way, the differentiation the center helps to bring about is not external to its inherent "nature." This is why the interlinked movements of centralization and decentralization always need and feed off each other, while still in conflict with each other.

Yet, by enabling effective control of very complex processes and series of action within a single structure, the telephone was conceived, like the telegraph had been, still largely through the analogy of the nervous system: the telephone took the shape of the nervous system of a modern industrial society,[139] its metallic nerve cells organizing the social body into a united organism capable of controlling itself. Thus John J. Carty, the chief engineer and subsequently the vice president of AT&T, combined the engineering worldview with the language of biological

136 D. H. Braymer, "New Police Telephone and Signal System," *Telephone Engineer* 3, (3), (March 1910): 127.

137 Pool, *Forecasting*, 73.

138 For the part of the collective movement in business, see Burlingame, *Engines of Democracy*, 130.

139 Casson, *History of the Telephone*, 191; Herbert Casson, "The Telephone as it is To-Day," *World's Work* 19, (April 1910): 12778. H. D. Stroud called it "the nervous system of the whole social structure" (H. D. Stroud, "Future Telephone Possibilities," *Telephone Engineer* 11, (1), (1914): 37). What is of interest is that while democracy was viewed by the progressives as depending on improved means of communication, not only were these means but also democracy itself often portrayed in terms of the same biological vocabulary. Thus it was presented as in analogy to the human body in which all cells were coordinated and pervaded by a single mind (Michael Pupin, *The New Reformation: From Physical to Spiritual Realities* (New York: Charles Scribner's Sons, 1927), 383–87. See also Tobey, *American Ideology*, 39).

analogy in his 1928 statement: "It is the function of the engineer to provide for the extension of the spoken word by means of electrical systems of intercommunication which will serve to connect the nervous system of each unit of society with all of the others."[140] Similarly, Frank Jewett, a young physicist who also became the vice president of AT&T in the following decade and the president of Bell Telephone Laboratories, was able to express the idea in 1936 that when there are no longer economic barriers to full usage of telephone lines, "society for the first time [will] have a complete nervous system."[141] Against this background it could be said that the telephone highlighted the concept of structured communication in the social organization, often interpreted in the light of organicist language, in at least three different respects, thus shaping the idea and limits of communication:

1) The significance of the local and regional purpose-built telephone systems cannot be underestimated, for it was essentially in relation to these diverse arrangements that the model of organized communication was formulated in this period and subsequently used more generally. In the United States Army, for example, three general classes of telephone system were operational in 1911. First, there was a system enabling instantaneous communication among all parts of an army post, analogous to a good regional telephone service. A second class of military communication system was designed to carry all communications relating to sea-coast defense, and the third was a system enabling communication between troops in the field and the command post or base of supplies.[142] The functional structure of the military communication systems naturally followed the same order of predetermined hierarchical relationships that was the basis of the military institution. As some of the first and foremost examples to exploit the communicational arrangement based on the telephone, they were to prepare the way for a general model for organized communication in large establishments outside the military. Yet, they were not the only source and setting for a rationally ordered communication system.

The telephone reorganized police communications following the example of the telegraph systems implemented in major metropolitan areas. Thus, the police communication system based on the telephone that provided immediate connection between police headquarters and the patrolman on the beat as well as among different departments and units, served as "the central nervous system of a highly

140 John Carty, "Ideals of the Engineer," *Bell Telephone Quarterly* 8, (2), (1928): 92. See also David Mindell, *Between Human and Machine: Feedback, Control, and Computing before Cybernetics* (Baltimore: Johns Hopkins University Press, 2002), 110.

141 Frank Jewett, "The Social Implications of Scientific Research in Electrical Communication," *Bell Telephone Quarterly* 15, (4), (1936): 216. Almost 40 years later it was still natural to think that "the telephone is our nerve-end to society" (John Brooks, *Telephone: The First Hundred Years* (New York: Harper & Row, 1975), 9).

142 Waldon Fawcett, "Telephones at United States Forts," *Telephone Engineer* 5, (4), (April 1911): 190. Lieutenant W. E. Palmer compared the US Army Signal Corps to the nervous system of the human body in his paper "The Function of the Signal Corps," *Telephone Engineer* 18, (1), (July 1917). For an elaborate history of the tactical use of the telephone, see David Woods, *A History of Tactical Communication Techniques* (New York: Arno Press, 1974 [1965]), 185–208.

integrated organism for the suppression and prevention of crime."[143] As early as in 1884, through the early police telephone and signal system in use in Chicago, each man "on patrol or watch" could be put "in communication in an instant with the subdivision to which he is attached, or, if that is necessary, with the police station of the district, and at the same time with the central station."[144] Similarly, the teletypewriter, which enabled almost instantaneous transmission of messages to numerous different places simultaneously, was introduced widely in business for it helped to rapidly coordinate the activities of factories, warehouses, and sales offices in different cities. But it also constituted the foundation for the extensive eight-state alarm system that provided the means for the police administration to coordinate the widely scattered state police posts, municipal police departments, and state motor vehicle and criminal identification bureaus.[145] Before the teletypewriter and the shortwave radio — a second major by-product of the telephone — "revolutionized" police practices,[146] telephonic communication fundamentally influenced the way in which these practices assumed a more efficient and rational structure, becoming the very "backbone of the [police] communication system."[147]

The New York Police Bureau used at least three different telephone communications systems as early as 1909, the aggregate effect of which was so considerable that it constituted a model to be followed. First, there were seven switchboards in New York "tied to and centralized upon the switchboard at the Manhattan headquarters [. . .]. Each switchboard is localized in its own borough, radiating to the courts, morgues, hospitals, humane and animal societies, police precincts, etc. The connections of each switchboard are brought direct to the central switchboard at police headquarters, Manhattan."[148] Thus, "while the central headquarters in Manhattan can connect with any precinct in any borough, [. . .] no local switchboard can communicate with a precinct outside of its own borough without going through the Manhattan switchboard."[149] This is how the communication structure of the police bureau was centralized on one location that organized the flows of interaction among the different parts and units of the system.

Second, in addition to and independently from this arrangement, there was a signal box system, consisting of 900 lines and 38 switchboards. It ran from precinct

143 V. A. Leonard, *Police Communication Systems* (University of California Press, Berkeley, 1938), 6. Cited in Pool, *Forecasting*, 3–4, 95.

144 George Prescott, *Bell's Electric Speaking Telephone: Its Invention, Construction, Application, Modification and History* (New York: D. Appleton & Co., 1884. Reprinted New York: Arno Press, 1972), 308–9. See also "A New Police Signal," *Scientific American*, (September 28, 1895).

145 Howard Hunter, "The Teletypewriter Helps the Police," *Bell Telephone Quarterly* 18, (3), (1939): 161–2; Willey and Rice, *Communication Agencies*, 146; David Woodbury, *Communication* (New York: Dodd, Mead & Co., 1931), 244; A. S. Benjamin, "Teletype: The Telephone Typewriter," *Telephone Engineer* 33, (12), (December 1929): 28–30.

146 Dilts, *Changing World*, 83.

147 V. A. Leonard, *Police Communication Systems* (University of California Press, Berkeley, 1938), 52. Cited in Pool, *Forecasting*, 96. See also V. A. Leonard, *The Police Communications System* (Springfield: Charles C. Thomas, 1970) for a more updated account. The telephone became so useful for police practices not only because of its adaptability to police uses but also because of its convenience for calls from citizens reporting crimes.

148 A. Rowsey, "Telephones in the New York Police Bureau," *Telephone Engineer* 2, (1), (July 1909): 6.

149 Ibid.

posts to local precinct houses only, and was used by patrolling policemen to make their required calls, "once an hour," to their local stations, as well as in emergency situations. The increase in this kind of controlled repetitive routine was part of the professionalization of public bureaucracies in this period, as already mentioned. Finally, there was a system of 43 leased lines running to the homes of administration and executive staffs so that should an emergency arise, "the executive staff would be hauled out of bed the first thing and told what to do and where to do it, while dressing."[150]

A combined police telephone and signal system enabled the chief of police "to get in touch with patrolmen in any section of the city in an emergency, and, further, makes it possible for the patrolmen to send special information to headquarters and to record it there."[151] These communication systems, complementing each other, indicated the growing importance of creating and organizing communication practices according to an institutionally executed, general scheme, and provided a feasible model for doing this for a number of social assemblages outside the field of policing. This model came down essentially to a predetermined communications relationship between the center and the terminals radiating from it, in which the latter reported locally gathered information to the center, which, after having processed the incoming information flows, sent uniform dispatches back to the terminal branches. One scheme of this kind was the Associated Press news service that was organized by telephone, following the arrangement based on the telegraph. Here information, gathered first all over the country by local reporters and news offices, was sent to the Associated Press head office in New York, from which, after filtering and editing, it was sent to subscribing members nationwide by telephone in one call. Thus, according to *Telephone Engineer* in 1911,

> [f]ixed call periods are set for both afternoon and evening, each period of half hour's length. The material to be sent in any given period is arranged and written with the idea of its taking just a half hour to read, and this, experience shows, allows some 1,200 words. The central telephone operator has the schedule of periods on file, and for the reading at nine o'clock she begins beforehand to connect the towns and the newspaper offices in them so that at nine o'clock she is able to notify the central office that the wire is ready. Then the news is read into a special instrument and in each newspaper office on the circuit an expert typist, wearing a head telephone receiver, takes down the news as it comes. This system is more direct than by telegraph, but it does not permit of the abbreviations and all the offices must wait while the sender repeats for any one that does not hear distinctly.[152]

150 Ibid, 7.

151 Braymer, "Police Telephone," 127.

152 "Associated Press Service by Telephone," *Telephone Engineer* 5, (5), (April 1911): 291. In the following year the system was described, in the same journal, from the specific perspective of two subscribing journals, which belonged to the circuit of Chicago's office: "At fixed hours each night the *Gazette* and *Mining Journal* are connected directly with the central office of the Associated Press in Chicago. An editor in the Associated Press office reads dispatches into the transmitter at the rate of approximately forty-five words a minute. [. . .] They are copied verbatim on a typewriter by the *Gazette's* operator and generally are ready for the linotype operators when they leave the typewriter, little revision being required. It is a common thing

In this scheme, which was subsequently superseded by the use of the teletypewriter in news reporting, the Taylorian division between "thinking" and "doing" found expression again. The central office had no firsthand experience of what was going on "in the field," and the communication line terminals had no final discretion concerning decisions to be made.

Thus it is clear that increasing use of the new systems of communication was inherently linked to existing social practices and discourses that, taken together, constituted, generally speaking, a framework for their reception and utilization. But the effective relationships ran also in the other direction, as we have seen, with technical structures being conducive to the formation of models and metaphors for representations of society. Local and regional communication systems, then, must be seen as having been created in the midst of diverse forces, and as always retaining an inherent relationship to current practices regarding the technicization of communication. Although these communication systems reflected the influence of Taylorism to varying degrees, at the same time they set in motion practices that reinterpreted classic scientific management in the light of local demands. Nevertheless, Taylorism undeniably formed the perspective for the telephone seen as a political technology.

2) The telephone strengthened the notion of the center as an operational necessity for the system as whole, because in contrast to the telegraph, architectural structure of the telephone system required central offices and switches in order to interconnect large numbers of telephone devices.[153] Whereas the telegraph interconnected locations, it was the mission of the telephone from the beginning to link people together, a capability made feasible by the central exchange, "the solar plexus of the telephone body,"[154] as John Mills has described vividly in 1925:

> The line of a telephone subscriber is like a hallway leading from the house, or office, where he has his telephone instrument, to the central office of this telephone city. And there he has a door opening onto all the national highways of speech. Through this telephone door and along the hallway, formed by the wires which connect his

for a dispatch of 200 words to be copied, headed and set in type all within twelve minutes" ("Newspaper Service by Telephone," *Telephone Engineer* 7, (3), (March 1912): 134). See also "The Associated Press: What It Is!," *Telegraph and Telephone Age* (January 16, 1916). In 1912, the *Telephone Herald* extended news reading over the phone to the entire subscribing public. In this experiment, "the subscriber does not read the Telephone Herald, but merely listens to it" (Arthur Colton, "The Telephone Newspaper: New Experiment in America," *Telephony* 62, (13), (1912): 391).

153 Neil Wasserman, *From Invention to Innovation: Long-Distance Telephone Transmission at the Turn of the Century* (Baltimore: Johns Hopkins University Press, 1985), 3–4. See also Stone, *Public Service*, 79.

154 Casson, *History of the Telephone*, 149. A contemporary telecommunications historian, in a breathless account, describes early telephone exchange in 1884 as "a system of telephone correspondence carried on through a central office, by means of wires radiating from it to the houses, offices, counting rooms, etc., of the subscribers to the exchange, a separate wire, as a general thing, being used for each subscriber, who is also provided with a list of all the other subscribers to the system, and when, at any time, desiring to communicate with any one of them, has merely to notify the central office, and the two corresponding wires are immediately connected, and direct correspondence is established" (Prescott, *Speaking Telephone*, 191).

> instrument to the central office, conversations can take place with anyone else who also has a door on the national highway.[155]

This and similar descriptions drew, often unknowingly, upon the outlines sketched out by Alexander Graham Bell. This is what he had envisioned in 1878 in a letter to the organizers of the Electric Telephone Company:

> In a similar manner [as in the network of gas-pipes and water-pipes], it is conceivable that cables of telephone wires could be laid underground, or suspended overhead, communicating by branch wires with private dwellings, country houses, shops, manufactories, &c., &c., uniting them through the main cable with a central office where the wire could be connected as desired, establishing direct communication between any two places in the city. [. . .] I believe in the future wires will unite the head offices of the Telephone Company in different cities, and a man in one part of the country may communicate by word of mouth with another in a different place.[156]

The telephone network was created in the division between the "route lines" (later termed "trunk lines"), on the one hand, and the "distribution lines," on the other. Route lines interconnected major centers, whereas distribution lines branched out from each center over a given area.[157] In practice, central exchanges appeared to users chiefly as the voice of telephone operators, as they — primarily young women — manually connected incoming calls to the numbers desired. This was why "the Central" became almost synonymous with the telephone for callers in early-twentieth-century America. It gave a conceptual form to the coherent nature of the telephone system in a way that was in harmony with both the contemporary sentiment searching for the possibility of scientific direction of society and the still effective organismic thinking. Thus the telephone system made up a "single unit from end to end," being "a single entity, almost a living organism which, if disturbed at one point, reacts throughout."[158]

Just as the brain was conceived to interpret signals sent to it by the central nervous system, it was the telephone exchange in which the "brain" of the telephone network was embodied.[159] Rudimentary exchanges were already in use in 1878, but they were not properly established until the technology developed in the next decade, when the first automatic switchboard was patented.[160] This made it pos-

155 Cited in Ellen Stern and Emily Gwathmey, *Once Upon a Telephone: An Illustrated Social History* (New York: Harcourt Brace & Co., 1994), 133.

156 Cited in John Kingsbury, *The Telephone and Telephone Exchanges: Their Invention and Development* (New York: Arno Press, 1972 [1915]), 90.

157 Kingsbury, *Telephone Exchanges*, 241.

158 Frank Jewett, "The Social Implications of Scientific Research in Electrical Communication," *Bell Telephone Quarterly* 15, (4), (1936): 214.

159 Casson, *History of the Telephone*, 143. In the 1920s, automatic switches were sometimes called "mechanical brains" (Mindell, *Human and Machine*, 131). Subsequently the computer assumed the role of electronic "brains"; e.g., David Sarnoff, "Electronic Revolution, Present and Future," *New York Times*, (September 30, 1956).

160 Kingsbury, *Telephone Exchanges*, 267, 394–400. See also Frederick Rhodes, *Beginnings of Telephony* (New York: Harper & Brothers, 1929), 147–66; Robert Chapuis, *100 Years of Telephone Switching (1878–1978):*

sible to dispense with manual exchanges, for now callers could contact numbers desired without the help of "Central." The automated telephone system encouraged more spontaneous and frequent use, which eventually established the telephone as a standard mode of communication in the industrial era. It is not incorrect to propose that in most of the organizations of the period, it was management that became viewed as a kind of switchboard. This is partly because management was commonly seen as representing the thinking agency in an organization, as opposed to the doing agency, not unlike the brain in the nervous system. In addition to this, however, management became a switchboard because it was the point in and through which key connections were made and that set the internal pattern for the communication structure. Thus, it determined the directions and routes of formal communication relationships between central and middle management and executive departments as well as created the space for interpersonal lines of communication. What management did was to organize communication within a hierarchical structure, yet it was already, of course, itself organized by and through the communication that continuously rearranged the social order, and with it, the corporate institution.

3) The telephone affected business practice and organization by providing a communication method for serving a combination of different units under the same corporate control.[161] It enabled an accelerating concentration of private power and an unparalleled accumulation of financial capital and social power that created new conditions for social and economic order. For this reason this period was also called "the Age of the Telephone."[162] It was not an invention like other, only superficially similar objects of use, because, in the words of Casson, "it is nothing less than the high-speed tool of civilization, gearing up the whole mechanism to more effective social service. *It is the symbol of national efficiency and cooperation.*"[163] It was established as a condition of modern social order alongside and related to new ideas in science and commerce, and these conditions seemed to assume the form of a strongly centralizing movement. It was Taylor's idea to combine this movement with scientific approach as a way of rationalizing the functioning of industrial organization in order to make production more efficient and to recast the institutional structure of society to meet the new *de facto* standards of performance introduced by technological development.

With the rise of Taylorism, the telephone was quickly harnessed to the service of rationalization, although, as already mentioned, it has been pointed out that the telephone did not play a significant role in the emerging formal communication system.[164] Yet its role should not be ignored either, as it clearly contributed to

Part I; Manual and Electromechanical Switching (1878–1960s) (Amsterdam: North-Holland Publishing Co., 1982).

161 Willey and Rice, *Communication Agencies*, 143.

162 Kenneth Boulding, *The Organizational Revolution: A Study in the Ethics of Economic Organization* (New York: Harper & Brothers, 1953), 207.

163 Casson, *History of the Telephone*, 238; emphasis in original.

164 Yates, *Control through Communication*, 21.

techniques exploited in scientific management and efficiency engineering. This was perhaps most instantly recognizable in train dispatching.[165] The telegraph had already established a practice of using telecommunications to coordinate the movements of the trains. Now, the telephone achieved the same ends more efficiently: "[T]he use of the telephone is so quick in every way and so much more flexible that the dispatcher gets far more detailed information of what each train is doing than was possible by the use of the telegraph."[166] It allowed more personal conversations, it was quick to use, and the weather did not make any difference to its operational dependability. But more important, it enabled detailed control not only over the movements of trains but also over what engine drivers were actually doing at any given moment: "[W]ith the telephone it is possible to arrange apparatus so that the superintendent can at any time supervise the actual work of train handling."[167] This was a decisive improvement in the overall coordination of such vast and complex organizational structures as the train companies were. Of course, the use of the telephone in personnel management did not remain confined to the train companies, but was spread in other industrial sectors as well, including the telephone services. Thus, a commentator could make the observation in 1916 that "the telephone makes [a telephone] official efficient, reliable, painstaking and ever prompt in the transaction of the business brought before him."[168] It provided a means to bind the workforce closer into part of a great industrial machinery.

This intention was also nicely embodied in a system in which the telephone was used for timekeeping purposes in the De Laval Separator Company. In this case, sub-foremen or clerks made a call to a central timekeeping department when a workman started and ended his job, the time taken in making this report apparently being less than 15 seconds.[169] This was a sophisticated system used for monitoring and "timing" the entire workforce of 650 men:

> When the sub-foreman starts a workman on a job he telephones to the timekeeper the class of work, man number, operation number and the number of pieces for the job just finished. Cards for production, re-work, construction, repair or expense (a distinctive color for each classification) are stacked up in front of the timekeeper, who, when he receives the message, takes the proper card, writes on it the information telephoned and opens the record with the calculagraph. [. . .] The timekeepers

165 See "The Railroad and the Telephone," *Telephone Engineer* 3, (1), (January 1910); K. W. Endres, "The Telephone for Train Dispatching," *Telephone Engineer* 3, (4), (April 1910); John Kick, "The Railway Telegraph Department," *Telephone Engineer* 5, (5), (May 1911); W. H. Clapp, "Telegraph or Telephone for Railroad Work?," *Telephone Engineer* 11, (4), (April 1914); "A Discussion on Railway Telephony," *Telephone Engineer* 12, (4), (October 1914); E. E. Trafton, "Train Dispatching by Automatic Telephones," *Telephone Engineer* 15, (6), (June 1916); W. A. Wolff, "Guiding Trains by Telephone," *Telephone Engineer* 16, (3), (September 1916).

166 "Train Orders by Telephone," *Telephone Engineer* 3, (3), (March 1910). See also Dilts, *Changing World*, 89.

167 "Train Orders by Telephone," *Telephone Engineer* 3, (3), (March 1910).

168 "Efficiency Engineering in the Telegraph Service," *Telegraph and Telephone Age* (January 1, 1916).

169 Edgar Snell, "Factory Cost Accounting by Telephone," *Telephone Engineer* 1, (5), (May 1909): 131–2.

> are required to call back over the telephone the figures received in each message before writing them down. So proficient have they become in recording the messages that, although fifteen to eighteen thousand cards are handled in a week the average number of mistakes of all kinds has been only five.[170]

Although this system was rather specialized, the telephone also played a part in general industrial intercommunication systems. This kind of system commonly consisted of

> a series of jacks, which are numbered to correspond with the different departments of the plant. A plug, hung on a conducting cord, is inserted into the desired jack and the pressing of a button immediately rings the signal bell in the corresponding department. [. . .] Every department in the plant may be reached by simply transferring the plug, without removing the receiver from the ear.[171]

In addition to improving internal communications, the telephone also made it easier to forge customer relationships: a growing proportion of businesses of all kinds were being transacted over the telephone. The New York Edison Company, for instance, capturing this fact in its 1911 pamphlet, stated that "[t]he public deals with us by telephone — we deal with ourselves by telephone."[172] Thus, the telephone began to assume an increasingly central role in the communicative structure of a business organization.

With the automation of telephony, its industrial role increased in terms of both technical solutions and the conceptual visions it supported. The automatic switchboard, it has been noted, was in a sense the prototype or conceptual model for the automatic factory that came to represent the whole industrial order in many cultural products of the 1930s, but which did not get a serious chance to be realized until the 1950s alongside the rise of the so-called system sciences.[173]

Taylorism can be seen as the culmination of the development in which social relations and processes were enshrined in a general law or method that was a precondition for the intelligent control of society. This law was dynamic and more sensitive to relationships between its elements and to changes to them, yet it was meant to fix them definitely in an immutable structure. The concept of a system coupled with the requirement for a central planning agency encapsulated expectations and requirements relating to adequate governance of how an assemblage consisting of diverse and constantly changing elements functioned. It opened up a field wherein it became possible to outline the essence of modern society in a situation in which old concepts did not seem to work any longer and the new ones were not yet established.

170 Ibid., 132.

171 O. N. Livermore, "Factory Telephones," *Telephone Engineer* 1, (2), (February 1909): 47.

172 Cited in Angus Hibbard, "Telephones in Central Station Control," *Telephone Engineer* 10, (1), (July 1913): 27.

173 Young, *Person to Person*, 82–3.

This alliance between the system and the center lent itself to thinking of automation as the general, uniting force, thus addressing this relationship in terms drawn from the technological realm. Automation supported the idea of the system having an intrinsic nature, on the one hand, but acknowledged also the need for regulating principle, on the other, and incorporated both into a consistent whole. Yet, before automation was introduced on a larger scale, and before cybernetics capitalized on the conceptual content of it, the idea had already been articulated in connection with the regulation of social processes. Essentially, the technocratic stream of progressive thought, in preparing to restructure society on a technical basis, visualized a system in which decisions "would be made automatically as a function of the industrial organization," as the technocracy historian William Akin put it.[174] Hence the aforementioned Wallace Donham, for instance, expressed in 1932 the idea that social control operations should be based on automated routines. He suggested creation of a mechanism "which will absorb the shocks [in society's adaptation to the environment] automatically."[175] This is where, very early on in this connection, the idea of bringing the center and the system together through controlled and repeated routines is raised, as exemplified in an automatic mechanism. This brought out the routine-based nature of automation, and in so doing illustrated at the same time the potential benefits of large-scale routinization of social processes within a controlled framework.

The concept of a centered system got firmly entrenched in social life and increasingly formed a point of departure for the structure of many organizations functioning in different social fields: the coordinating center and the sectional units grouped under it began to constitute the model on which especially commercial and political organizations seemed to depend. The center is not to be understood, however, only in its structural sense as part of an organizational model, since it was associated with a whole set of practices that permeated the social order. This was related to a process in which bureaucratic control, having taken shape in the late nineteenth century, brought the traditionally decentralized and predominantly locally administered American society into its sphere. During the course of this, social practices began to be rationalized through professional national bureaucracies (the tax authorities, the police and fire departments, for example), although the role of public bureaucracies remained limited in comparison with that in many European nation states. In any case, the concept of a rational and efficient whole, which took a distinct shape from the operational models based on communication systems and other technical structures, largely provided the basis for discourses and conceptual forms applied to thought on society and a starting point for the gradual but resolute process of formalization of communication beyond the realm of engineering.

This viewpoint allows us to go one step further and delineate schematically the relationship between centralized power and the formalization of communication: the center could be conceived of as seeking the optimal elimination of multiplicity

174 Akin, *Technocracy*, 79.
175 Donham, *Unforeseen*, 34.

or noise — a concept that subsequently came to be exploited in cybernetics and information theory — through the rationalization of innate system functionality, as it were. This aim, reflecting and using scientific method, constituted the conceptual and institutional terrain in which communication began to be formalized. Though communication is not, of course, reducible to the former alone, for it has its own singular histories, thought about it was anyway moving into contexts in which science figured as a predominating measure of the truth of it. Communication was exposed to the objective norm in order to eliminate anything merely incidental to it and to reveal its essential core: what was of interest was only its hidden, necessary logic as knowledge of this was assumed to give access to the desired organizational rearrangements, thus improving general system functionality. But centralization was itself one of the most outstanding outcomes of organizational transformation during the period in question. Therefore it is in no way independent of the methodological explanation of communication. On the contrary, whereas institutionalized scientific practices determined the true nature of communication by separating noise from it, it was precisely centralization that carried out this separation in terms of improved efficiency. Another way to describe this mutual relationship is to say that as institutional requirements necessitated a center, it was the systematic examination and regulation of communication — the task imposed by the method as encapsulated by the center — that called for the institutional reform. Thus, while science forged a relationship with the truth of communication, the center built one with the efficiency of communication, because for the latter the separation of what was random from what was necessary was always more a question of control rather than of knowledge, although these aspects, it is true, cannot be disentangled in a clear-cut way from each other.

Telephonic Communicativity

I have attempted to show how a new relationship was gradually forged between "communication" (as both a concept and a process) and "telecommunication" (as the set of interrelationships among distinct technologies, institutions, and discourses) during the period from about the turn of the century until World War II. As already discussed, two developments in particular came to the fore and opened up a new relationship with communicativity: first, scientific management became the political technology of bureaucracy, and second, the telephone was established as a prerequisite for the modern community, the very embodiment of advanced communication. Let me now take a closer look at how this communicativity, as an alliance among technologies, discourses, and practices, unified diverse communicative events under the same whole: a whole that was serial or mechanical in nature, on the one hand, without being, however, independent of these events, on the other.[176]

176 Although the image of a consistent communicative whole had definitely started to disintegrate, it still remained an important framework for thought. Images of the telephone system as having a coherent structure, despite the apparent difficulties in viewing it as a whole, abounded. Frank Jewett, in a speech

Chester I. Barnard, in his seminal book *The Functions of the Executive* published in 1938, developed the idea that all successful organizations depended on cooperation, and that it was the task of managers to create it through communication. Barnard, whose background was in telecommunications, subordinated communication to a social whole whose function it became.[177] Hence he operated within the tradition in which communication had been conceived as both an object and a tool of administration. Yet he systematically promoted the constitutive characteristic of intercommunication as part of a theory of formal organization. Barnard explicitly described the system of communication as an essential part of any organization: "[C]ommunication technique," according to him, "shapes the form of and the internal economy of organization."[178] Moreover, the system of communication was "a primary or essential continuing problem of a formal organization. Every other practical question of effectiveness or efficiency — that is, of the factors of survival — depends upon it."[179] He made communication an explicit object of analysis during a time when its quintessential role in organizations was realized but not yet generally studied, contributing thus not only to the theoretical understanding of communication but also to the institutionalization of the theory of organization. "In an exhaustive theory of organization," he stated, "communication would occupy a central place, because the structure, extensiveness, and scope of organization are almost entirely determined by communication techniques."[180] Barnard thus made communication the focal point of theories aimed at understanding the functioning and dynamics of modern organizations.

As communication became conceptualized, the formal, hierarchical structure of communication was particularly emphasized, as a large part of communication in organizations was informal.[181] Accordingly, it was the task of the manager to establish an objective hierarchy of lines of communication out of a sea of informal, spontaneous, and habitual communicative relationships. This required lines of communication to be defined, formal channels between the manager and every member of the organization to be determined, and conflicting communication be avoided by establishing procedures concerning the use of the communication channels. This idea of a formal or objective ordering of communication would not, of course, have been possible without the contribution of Taylorism, from which it largely received its theoretical grounds.

delivered in 1939, dealt with the question of visualizing the telephone network: "In considering the evolution of telephony, we are dealing with a structure that does not lend itself to pictorial representation. But if the reader could be transported rapidly, magic-carpet-wise, out across the continent 3,000 miles or so, thus enabling him to visualize what a complex thing is this creation we are considering, he might acquire at least a superficial conception of its elaborateness and seeming fragility" (Frank Jewett, "A Quarter Century of Transcontinental Telephony," in *The Telephone: An Historical Anthology*, ed. George Shiers (New York: Arno Press, 1977 [1940]), 3).

177 Chester Barnard, *The Functions of the Executive* (30th anniversary edn), (Cambridge, MA: Harvard University Press, 1968 [1938]), 49. Barnard joined AT&T in 1909 and became the president of New Jersey Bell in 1927.

178 Barnard, *Functions*, 90.

179 Ibid., 175.

180 Ibid., 91.

181 Ibid., 109, 175, 217.

As important as the goal, however, was the idea of the whole within which this ordering took place, and this whole resembled closely the nervous system, with the manager being the center through which all relevant information flowed. As in the nervous system, it was the function of this center of communication "to translate incoming communications [. . .] into outgoing communications."[182] Thus Barnard could claim that

> the functions with which we are concerned are like those of the nervous system, including the brain, in relation to the rest of the body. It exists to maintain the bodily system by directing those actions which are necessary more effectively to adjust to the environment, but it can hardly be said to manage the body, a large part of whose functions are independent of it and upon which it in turn depends.[183]

The primary executive function, according to him, remained "the establishment and maintenance of the system of communication."[184] Through Barnard, the common view that the structure of communication of a given system could be used as a means for advancing its efficiency was given theoretical legitimacy. Moreover, the fact that certain communication systems had been taken as conceptual models on the basis of which complex social and industrial systems could be both understood and directed appeared to gain more theoretical justification.

Now, it would surely be an overstatement to attribute some kind of privileged status to these communication systems with respect to a rationalization process that was not limited to the business realm. Yet there has always been a connection between these kinds of systems and the existence of and as a community. Comments such as "the civilization of a country seems to advance in proportion to the development and extension of its communication systems," as a commentator remarked in 1929,[185] were indicative of this strongly felt relationship. Therefore it is not far-fetched to assume that communication systems were a formative part of the great rationalization movement, especially as it was at the local level where the main focus of the reform lay.[186] Thus they can be said to have prepared the ground, for their part, for generalization of the substitution of a technique for endless negotiation through the gradual institutionalization, growing, and merging of diverse local systems into a more or less unified view of an organized whole.

These local systems, combined with the rational planning discourse and available communications technology, formed an essential part of the history of the formalization of communication and its generalization within society — a line of development the organizing of which was associated rightly with the seminal contribution of Taylor. New Deal liberalism, borrowing from the progressive

182 Ibid., 178.

183 Ibid., 216–17.

184 Ibid., 218, 226.

185 Ray Blain, "Communication Is an Element in National Progress," *Telephone Engineer* 33, (March 3, 1929): 24.

186 Wilson, *National Planning*, 26; Alan Brinkley, *Liberalism and Its Discontents* (Cambridge, MA: Harvard University Press, 1998), 56.

demand for greater governmental regulation of society, and managerial liberalism with its focus on avoiding state bureaucracies by means of efficient corporate management and coordination, embodied this generalization of the belief in rational planning on a national scale with its emphasis on the politically neutral science of the administrative process.[187] President Hoover appointed the Research Committee on Social Trends in 1929 in order to bring academic research to the service of the federal government, and this confidence in objective knowledge also guided President Franklin D. Roosevelt in the establishment of the Committee on Administrative Management in 1936.[188] Given the particular task of applying scientific expertise to diverse social problems, the New Deal programs experimented with and created forms of scientific governance, the idea of which proved fairly resilient in twentieth-century Western political history.

Importantly, in its aim to set up countervailing centers of power against big business so as to protect the interests of the public, the New Deal also established public arenas for mutual negotiation and reconciliation between opposing interests, reflecting what has been termed "interest-group democracy."[189] Through the National Recovery Administration and its "codes of fair competition" for "stabilizing" business practices, New Dealers attempted to rationalize the major industrial sectors of the American economy. With the National Recovery Administration and other industrial recovery agencies and schemes they sought to forge government–business co-operation mechanisms in the interests of national planning and of national unity, although in the early stages little may have been done to speed recovery.[190] In any case, the shifting of the grounds of democracy from ethical to technical considerations[191] promoted a technical outlook in the political discourse.

In line with the goals of the New Deal, the 1934 Communications Act shaped the institutional framework for national telecommunication regulation in the form of Federal Communications Commission (FCC), thus creating a specialized agency for the communications industry. President Roosevelt stressed the need for

187 Akin, *Technocracy*, 2; Wilson, *National Planning*, 35; Alan Dawley, *Struggles for Justice: Social Responsibility and the Liberal State* (Cambridge, MA: Belknap Press of Harvard University Press, 1991), 64, 142, 163, 295; Horwitz, *Regulatory Reform*, 25–6.

188 Karl, *Executive Reorganization*, 37–9, 206–8, 229, 243–4. However, this is not to claim that there existed any strong intellectual or political continuity between the polities of the 1920s and the New Deal, or any inherent unity in the New Deal programs for that matter. As historians have argued, although building on the managerial experiments of the 1920s around federal institutions, the New Deal did not have the capability to synthesize public and private goals into a consistent political program. Thus the interpretations that emphasize the business-driven, prescriptive visions and models in politics seem to ignore the disorganization of the business and the fragmented responsibilities and resources of the state (e.g., Colin Gordon, *New Deals: Business, Labor, and Politics in America, 1920–1935* (Cambridge: Cambridge University Press, 1994), 29–33, 166, 281).

189 Whereas Professor William Leuchtenburg used the term in 1963 (William Leuchtenburg, *Franklin D. Roosevelt and the New Deal, 1932–1940* (New York: Harper & Row, 1963), 87), the term "interest group liberalism" was introduced by Professor Theodore J. Lowi in 1969 (see Hanson, *Democratic Imagination*, 280).

190 Leuchtenburg, *Roosevelt*, 64–71, 84–8; Anthony Campagna, *U.S. National Economic Policy 1917–1985* (New York: Praeger, 1987), 117; Brinkley, *Liberalism*, 17–62.

191 Hanson, *Democratic Imagination*, 292.

central regulation when he submitted a communications bill to Congress, stating that he had "long felt that for the sake of clarity and effectiveness the relationship of the Federal Government to [. . .] 'utilities' should be divided into three fields: transportation, power and communications. [. . .] In the field of communications, however, there is today no single agency charged with broad authority."[192] The Act could be viewed as embodying the striving, which characterized New Deal reforms, for unifying and centralizing government decision-making authority over a field of particular social activities. Thus one of the main motivating forces was to defend the "public interest," a phrase that appeared throughout the Communications Act as the ultimate measure for the different regulatory functions to be applied.[193] The new FCC divested the Interstate Commerce Commission of telecommunications jurisdiction and replaced the old interest in railroads with a more determined focus on questions of communication, and especially of telephony.[194] If, after the 1910 extension of the Interstate Commerce Act to include communications, there were still provisions that did not apply to its common carriers, the Communications Act of 1934 corrected this by extending the entire scope of the regulatory scheme to them.[195] Now all communications problems could be referred to a single office, which divided them into three separate categories: broadcasting, telegraph, and telephone. Significantly, the effects of communications came to be included in a sphere of systematic examination, and telecommunications policy became an institutionalized concern of American politics.

It is true, of course, that the tendency toward increasing rationalization by way of centralized decision-making cannot be reduced to a few managerial tenets, for these ideas were rather representations of a larger state of ferment wherein social order was constantly rearticulated. It was during this process also that conceptualizations concerning society itself became reformulated. Whereas in the late nineteenth century it was common to understand society in terms of a biological organism, the need to reshape theories concerning the underlying nature of

192 Included in US House of Representatives, *Report to Accompany S. 3285* (June 1, 1934). Cited in Stone, *Public Service*, 276. See also "Communications Report to Commerce Committee," *Telephone Engineer* 38, (2), (February 1934).

193 Ronald Cass, "Titles IV, V, and VI: Review, Enforcement, and Power under the Communications Act of 1934; Choice and Change in Institutional Design," in *A Legislative History of the Communications Act of 1934*, ed. Max Paglin (New York: Oxford University Press, 1989), 79; Robinson, "Communications Act," 14; Horwitz, *Regulatory Reform*, 25–6. See also House Doc. No. 340, 76th Cong., 1st Sess., (1939).

194 Brooks, *Telephone*, 196; Lindley, *Impact of the Telegraph*, 131–2; Mueller, *Universal Service*, 6, 157; Manley Irwin, *Telecommunications America: Markets without Boundaries* (Westport, CT: Quorum Books, 1984), 5, 27.

195 Kenneth Cox and William Byrnes, "Title II: The Common Carrier Provisions; A Product of Evolutionary Development," in *A Legislative History of the Communications Act of 1934*, ed. Max Paglin (New York: Oxford University Press, 1989), 30. This reflected the shift from what started as a general regulation of the economy, motivated by concerns about economic instability, to more industry-specific controls, as in the communications field (Horwitz, *Regulatory Reform*, 9, 65–75). By the end of the 1940s, however, there were still demands for more comprehensive efforts to develop "a total national communications policy"; see President's Communications Policy Board's 1951 report *Telecommunications: A Program for Progress; A Report by the President's Communications Policy Board*. In *Documents in American Telecommunications Policy: Volume Two*, ed. John Kittross (New York: Arno Press, 1977).

society became evident. While scientists examining the laws of communication and communications systems have been inspired by the structures of living bodies since the introduction of telecommunication,[196] it was precisely this structure that became the target of systematization in the early twentieth century — or, as the political scientist and philosopher Arthur Bentley put it in 1908, "in living beings there is no function that is not systematized."[197] Though the analogy of a nervous system remained in use, it was now conceived of in an altered, more technological context. The sequence of communicative processes in which the consequences of human action formed complex, unexpected, and impersonal chains of effect called into question the idea of a natural, harmonious social order. Instead it suggested a vision of an intricate system, the mutually interacting parts of which seemed to form a functioning machine. It was communication among the component parts that made up a system out of an apparently heterogeneous whole, and it was in this system, or rather machine, that the social organism assumed its new, industrial identity.

As it became evident that the parts and processes of society were linked together like the components of a machine, biological analogies were gradually replaced by mechanical ones, or at least were interpreted in the light of them.[198] This interpretation gained ground concurrently with the spread of the mechanism paradigm in the social sciences: thus, whereas "psychology is the science of the *mechanism* of the social process," wrote sociologist Albion Small in 1903, "sociology is the science of the mechanism of the *social process*."[199] The image of society as an enormous, "highly integrated machine" or "mechanism"[200] made possible a new understanding of communication as a hierarchically organized system rather than as a set of "organically" developing relationships. Thus, distinct communicative phenomena, even though superficially independent of one another, could be understood as ultimately forming a dynamic organization with each of its components dovetailing with one another. This is because "any mechanism which tends to strengthen the bonds between the various units of the industrial system would tend to make the system more like a functioning whole."[201] Hence, for instance, according to the aforementioned *Technocracy Study Course*, "the general function

196 Otis, *Networking*, 4.

197 Bentley, *Process*, 285.

198 See Elsner, *Technocrats*, 134; Haber, *Efficiency and Uplift*, 119; Wiebe, *Search of Order*, 146; Hart, *Forged Consensus*, 66; Israel, *Machine Shop*, 42; Marcus and Segal, *Technology in America*, 140. Of course, there was already a conceptual differentiation between "organism" and "mechanism" by the end of the eighteenth century, as Carl Schmitt explicated in a book originally published in 1936. According to him, the philosophy of German idealism drained the image of "mechanism" of all living characteristics, rendering it an inanimate entity without any mythical quality. This represented a deviation from Hobbes and his contemporaries, who considered both products of the highest human creativity. Thus, according to this interpretation, mechanism still carried a thoroughly mythical meaning for Hobbes (Carl Schmitt, *The Leviathan in the State Theory of Thomas Hobbes: Meaning and Failure of a Political Symbol*, trans. George Schwab and Erna Hilfstein (Westport, CT: Greenwood Press, 1996), 41).

199 Cited in Sklansky, *Soul's Economy*, 172.

200 Frank Arkright, *The ABC of Technocracy* (London: Hamish Hamilton, 1933), 81. Howard Scott described the United States as "a huge and intricate machine" (Scott, "Price System," 139).

201 Rexford Tugwell, *Industry's Coming of Age* (New York: Harcourt, Brace & Co., 1927), 170–1.

of communications [. . .] — mail, telegraph, telephone, and radio — automatically constitutes a functional unit."[202] In social order, each event and instance had its own place and function: the concept of the system became the absolute measure of the organization in late-nineteenth- and early-twentieth-century American society.[203] The ultimate structuring principle of the world was now recognized as hierarchical through the intervention of technical systems; and no pains were spared to determine the composition of this hierarchy, to standardize the elements belonging to it, and to designate the inherent place of each element and establish their interrelationships. Social relations were thus arranged in a system of hierarchically organized components.

Systemicity was not only a neutral, intrinsic principle of the structure of society but also something desirable in itself. Mechanisms that increased systemicity by multiplying interconnections among components of the system were seen as contributing to the functionality of the whole. Improvement of communications was naturally the key goal, as already discussed. Rexford Tugwell, for instance, a social planner who acquired fame particularly during the New Deal, regarded communication, against this background, along with transportation and exchange markets, as the nerve and coordinating center of the economic organism. As it develops, "the organism functions in a more complete fashion, more intelligibly, more as a whole. [. . .] When, then, we link a continent with telegraphy and especially telephony [. . .] we quicken the nervous system."[204] Insofar as division of labor, specialization of function, standardization of technique, and rationalization of operations became the basis for industrial production,[205] communication became a system that was definable in mechanistic rather than biological terms and prepared the ground for models inspired by the mechanical automaton.

The crux of this changed conception resided in the insight that society had an intrinsic functionality at a systemic level. This underlined the need for an overall centralized control over interlinked parts and processes, which society, as opposed to the biological body, did not posses. The biological analogy proved unsatisfactory because at the time there was no "brain for our social organism."[206] Machinery with adequate connections existed, but the social body seemed not to have a central organ corresponding to the physical facilities in order for it to be feasibly likened to a biological body. It was, however, exactly this kind of coordinating organ that was desired, for the social organism "must have a mind."[207] The idea of a center implied an unprecedented importance in the operational position of, in the words of Donham, a "central thinking agency" as it would "anticipate the slow trial and error of organic evolution."[208] In this line of thinking, the notion of

202 *Technocracy Study Course*, (1934), 221.

203 See Peter Miller and Ted O'Leary, "Hierarchies and American Ideals, 1900–1940," *Academy of Management Review*, 14, (2), (1989): 257; Marcus and Segal, *Technology in America*, 133–40.

204 Tugwell, *Coming of Age*, 172.

205 Marcus and Segal, *Technology in America*, 140.

206 Donham, *Unforeseen*, 49.

207 Ibid., 31.

208 Ibid., 49.

a rational center was strengthened, but was now seen as an agency for maintaining the "intrinsic" functionality of the system. The center did not itself generate this functionality but could only use precepts based on social laws that social research would discover over time. Hence, the analogy of a machine introduced the idea of system rationality into concepts of society in which the theme of a coordinating center figured ever more strongly.

The roots of this concept are not remote from the field studied by the quoted writers, for they can, in the main, be traced back to centralized systems and various related discursive practices based on telecommunications and other technical structures. The telephone as the main form of electric communication seemed to assume a special role in this connection, because it provided an operational model not only for modern communication but also for many industrial institutions. These dimensions were, of course, inherently intertwined: industrial society requires and produces communicative practices corresponding its needs and communication in its modern forms provides the basis for the prevailing social order. What is especially noteworthy is that communication as the precondition of industrial production becomes at the same time a prerequisite for thinking about society, because when speaking about contemporary societies we have to presuppose the role of the highly developed communication systems that make these societies possible. And it is exactly from the sphere of communication that not only the technical means but also largely the still extant forms for thinking about this relationship have emerged.

Thus according to the aforementioned Frank Jewett, for example, the network of various communication channels constitutes, in practice, one machine. This was an idea that was later generalized in cybernetic thought through an analogy combining the concept of the nervous system with the computing machine:

> The communication channels which interconnect the switchboards within a great metropolitan area, and the long distance lines which criss-cross the country and form the interconnecting network between cities, towns and hamlets, are in the last analysis merely the extended wires which link together parts of a single switching machine. [. . .] [T]hey are functionally no different in their purpose from many of the wires which connect together different parts of a switchboard within the same central office.[209]

While the idea of communication as a machine created a platform for planning and formalization experiments and gradual consolidation of certain modes of creating organization out of disorganization, what constituted the whole within which this all took place was, as shown, none other than the nervous system. "[Thousands of central office switchboards] are merely parts of a single complex

209 Frank Jewett, "The Telephone Switchboard: Fifty Years of History," *Bell Telephone Quarterly*, 8, (1928): 153.

organism,"[210] constituting the "nervous system of the whole social structure."[211] It gave a common name to the diverse systems and processes of organization, as well as the structure through which the organization could be realized. Hence, it was the form of the organization — that is, a conceptual means for conceiving the field of relationships as a whole — as well as what was organized in the course of this that was this whole itself. The nervous system, thus, served as the foundation of these projects that gave rise to the organized system of communication.

Politics as a formalization of social functions emerged from the possibility of thinking about modern Western community through the machine metaphor. In this metaphor thinking about commonality was expressed as a demand for adjustment and regulation rather than as a matter of negotiation around different interests. It is important to notice that the notion of communication as an ordering technology, as proposed by the technocrats and later elaborated on by the cyberneticians, was firmly entrenched in the thinking that characterizes the history of communication as the objective and means of administration. It suffices to quote a renowned telecommunication theorist from the late 1950s: "Communication is a social function. [. . .] [T]he whole, the organization or organism, possesses a structure which is describable as a set of *rules*, and this structure, the rules, may remain unchanged as the individual members or elements are changed. By the possession of this structure the whole organization may be better adapted or better fitted for some goal-seeking activity."[212] The machine metaphor becoming established through efficiency rhetoric and especially through scientific management did not appear in contemporary discourses as a distinct conceptual construction but rather as a result of changes in the structure of production in an industrializing society and as a response to the problems of such a society. It formed, hence, the boundary within which issues pertaining to reorganization of the social system were conceptualized and means of achieving it were addressed.

The machine-likeness of communication was reflected in the concept of a communicative whole that consisted of interconnected parts differing functionally from one another. This manifested itself after Taylor mainly in the separation of planning from doing, with the communicative relationship between them consisting largely of instructions from the former domain to the latter one. This arrangement established formal and unequivocal chains of command and by so doing set free the potential for more efficient functioning of the whole. It emphasized the functional separation of action as the basis of communication: a whole should be as adapted to its purpose and program as possible. Division of tasks was implemented through the way the mechanical assemblage standardized goals, specialized roles, routinized practices, and regularized communication.

New means of communication, coupled with the related discourses, changed the tasks and functions of communicative practices into part of a larger communicative whole, with social and institutional actors influencing the way this whole

210 Ibid., 151–2.
211 Stroud, "Telephone Possibilities," 37.
212 Colin Cherry, *On Human Communication* (3rd edn), (Cambridge: MIT Press, 1980 [1957]), 6.

was continuously reconstructed. After Taylor had separated thinking from doing, it became clear during the development of modern management that their functional relationship was far more complicated than he had given occasion to think. This understanding of the complexity of an organization was interwoven with the vision provided by the analogy of the nervous system that now had been given a mechanical interpretation. The center of a hierarchy did not only assign functional departments according to some pre-thought plan but also in the same way that the functions of a body were largely independent with respect to the brain, so functional departments in modern business organizations assumed a fairly self-governing role. Of course, they remained accountable to management but their interrelationship could be characterized better as a complex structure, directed through intermediate levels of regulation, than as a straightforward relationship of command. What is more, the center presupposed this functional relationship as a condition of its existence.[213] It was this terrain that cyberneticians came to work on, thus becoming involved in the ontological history of communication.

213 See Barnard, *Functions*, 217.

4

Logical Assemblages

After World War II a completely new form of communicative order that emerged as a result of the interweaving of two fields or processes began to take shape. First, the systems-scientific approach spread and became common, and second, information and communication technology developed and came to be seen as a necessity. These two processes were naturally not separate but, on the contrary, were mutually articulated in a symbiotic relationship. Systems sciences and especially cybernetics and information theory were established as the theoretical sources for various systems of communication and computation,[1] and computers, on the other hand, enabled the realization of systems-scientific applications. Together they engendered a fairly consistent form of communicativity that did not start to weaken until the 1970s. "Communication" and its position as an organizing factor in key social processes was given a new interpretation, and this novel way of thinking was extended to the general understanding of social agency. The significance of emergent digital computing technology was absolutely crucial to the ontology of communication. This technology, coupled with new fields of knowledge becoming institutionalized, made society a huge information structure. As a result, the totality within which communication and central social processes were thought about changed from "machine" to "information system," rendering it a far more exact entity than previously.

Communication becoming something more and more precisely definable in numerical values was, of course, not just of interest in academic discourses; it did not arise solely out of them and was not an object for them only. It was, above all, the result of practical needs, stemming largely from the development of telecommunications in the context of modernity. Physical implementations of telecommunication posed questions about communication that were given appropriate formulations in their fields by telephone engineers and mathematicians. Behind it all was chiefly a need to make communication economically profitable by using as few lines as possible to transmit the greatest possible quantity of information.[2] The quantification of communication is essential to the reliability of a

1 Steve Heims, *The Cybernetics Group* (Cambridge: MIT Press, 1991), 282.

2 As Jacques Lacan put it: "The Bell Telephone Company needed to economise, that is to say, to pass the greatest possible number of communications down one single wire. In a country as vast as the United States, it is very important to save on a few wires, and to get the inanities which generally travel by this kind of transmission apparatus to pass down the smallest possible number of wires. That is where the quantification of communication started" (Jacques Lacan, "The Circuit," in *The Seminar of Jacques Lacan*, ed.

communication system, and becomes problematic, according to contemporary critique, only when communication, outside the engineering field, is considered primarily in quantitative terms.

As has been shown already, it was precisely telecommunication systems that mainly constituted the environment in and through which communication assumed a distinct identity of its own. Since the very beginning of electric communication, questions relating to the speed and capacity of transmission of "intelligence," as it was commonly called — anticipating and paving way for the concept of "information" — had constituted a basis for the quantification of communication. This was subsequently accomplished through the concept of "the quantity of information" that was useful in analyses focusing on the problems of reliable and rapid transmission of signals, and came to be regarded mainly as a technical question of modulation.[3] This is how communication was established as an object of systematic, institutionalized investigation by professional scientists and technicians, as something that could be described primarily in numerical terms.

This quantification of communication became possible only in the social historical phase in which it had been given a market value — in other words, in the stage in which it was incorporated into the capitalist production process and made a commodity. Communication, however, has never been external to this process, like something that would have been caught in the market mechanism only at a certain stage of development as a resource. On the contrary, and as noted earlier, communication, especially in the form of telecommunication, was the very force, linked to a number of social processes, which revolutionized the social conditions of economic activity and paved the way for a new, industrial social order.[4] Thus, the quantification of communication cannot be disentangled from its own history and reality; its quantification is not external to communication in itself — in its modern mode — but is instead an inseparable part of its own historical development.

Yet it is possible to draw a distinction between communication and telecommunication and propose that the fate of the former is not reducible to that of the latter; but two noteworthy points relate to this proposal. First, it can be noted that telecommunication largely establishes nowadays the social environment within which communication is thought about. It forms the context in and through which a communicative relationship is accounted a social relationship; in other

Jacques-Alain Miller (New York: Norton, 1988), 79). In 1848, a contemporary commentator put the same issue with regard to the telegraph in the form of a simple question: "How can the greatest amount of intelligence be communicated in the fewest words?" ("Influence of the Telegraph upon Literature," *Democratic Review* 22, (May 1848): 409).

3 Colin Cherry, *On Human Communication* (3rd edn), (Cambridge: MIT Press, 1980 [1957]), 41, 43; M. E. Maron, "On Cybernetics, Information Processing, and Thinking," in *Cybernetics of the Nervous System*, ed. Norbert Wiener and J. P. Schadé, vol. 17 in *Progress in Brain Research* (Amsterdam: Elsavier, 1965), 120.

4 Again, this is not to say that telecommunication would have been the only or primary agent of change. Rather, seen from the viewpoint of this book, it provides perhaps one of the best perspectives to the dynamics of social change.

words, it organizes communication as something socially significant and collectively constitutive. Moreover, there is a danger here in postulating an original, privileged, and, with respect to its particular historical realizations, independent form of existence for communication; it is anyway impossible for communication is nothing other than the passage of history in which everyday social relations are expressed and thought about.

Given this, communication can be differentiated in principle from the changeable forms of its historical manifestation, as noted earlier, even though it does not differ from those in any way, due to the fact that social relationships and orders articulated through them constitute in any given situation what communication *is*. This differentiation is not a determination of any meaning external to communication, but is not based, either, on any existing difference: it is nothing substantial in itself. The point where the distinction between communication and its historical modes can be drawn is in communicative existence itself. There is nothing being-like or substantial in this communication, no ideal source of significance that would enable its expression in positive terms. Though the distinction is not pre-given, it is always made — and it is important to constantly maintain this separation between the constitutive realm of communication and its historical manifestations. However, what this also means is that differentiation is not dependent on individual choice but is instead something derived from communicative existence itself.[5] Thus we can conceive that the development of communication into something that can be dealt with by formal means and expressed in quantitative terms is indistinguishable from our modern vicissitudes.

Projects in and through which attempts were and still are being made to formalize communication are compelled to postulate some form of authenticity, typically in the form of a communication system understood as a totality. The modern history of communication assumed forms in which a certain founding logic presented both the possibility and the limits of its realization. Here this logic is understood as a law: it is the comprehension and treatment of communication in the context of a more general order that determines in which forms they can occur. It goes without saying that, because communication always connects with discontinuous processes and practices, it cannot be simplified into a single form or disposition, not even into the most influential and unchallengeable one. However, what makes communication understood as a law of paramount importance here is that it reveals what characterizes the whole of Western culture in its modern form. But, as was noted earlier, one has to keep in mind that communication itself was both a starting point and an object of this law, which naturally was not restricted to the laboratories of telecommunications engineers but became an integral part of modernity. In this process, communication became an instrumental means of organizing the social realm as well as of its self-organization. If the form of this self-organization was more than ever before based on law, now

5 Jean-Luc Nancy has aptly elaborated this theme with reference to Heidegger in his essay "The Decision of Existence." See Jean-Luc Nancy, *The Birth of Presence* (Stanford: Stanford University Press, 1993).

the law appropriated qualities that changed it from being a stable and unchanging principle into a self-governing autonomous mechanism.

The social shaping power of communication could be seen, for example, in telecommunications' founding of systems that enabled different social practices to be accomplished in a predetermined and routinized way. What this meant exactly can be illustrated by referring to the numerous but formally similar implementations of different systems, ranging from intraorganizational communication systems, including vast governmental configurations, to nationwide mass communication schemes. Communications technology contributed to differentiating responsibilities and positions, as well as dividing tasks into processes that were amenable to governance and control. In this way communication, through its own existence — that is, in the process of its own formalization and not in some external domain of reality — prepared the ground not only for objectification of communication but also for making social processes into the object of formalization.

In what follows, I will examine the practices within which communication was redefined as a distinct subject and means of governance in the age of the computer. In particular, I am focusing on how the conception of communication as an automatic system began to steer thoughts about society during the Cold War period. Because of the dominant positions of cybernetics and information theory as points of intersection of theoretical, technical, and administrative lines of discussion and practice during this period,[6] the first part of the chapter will investigate formation of the concept of communication based on these theories, whereas the second part is devoted to the relationships these ideas opened up with thinking about "the social" through mutually articulated theoretical discourses and their practical applications.

Automating Communication

The engineering and managerial models in and through which communication was concretized into rational systems and which were first formulated at the beginning of the progressive era, found their highest expression in the 1950s and 1960s.[7] It was the rise of the so-called system sciences with their mathematical

6 When arguing that the sciences had assumed the role traditionally held by philosophy, Martin Heidegger, in his famous *Spiegel* interview *Nur noch ein Gott kann uns retten*, originally published in 1976, named "cybernetics" as the very field of knowledge and practice that was about to take the place of philosophy (Martin Heidegger, *Antwort: Martin Heidegger im Gespräch*, ed. Günther Neske and Emil Kettering (Pfullinger: Neske, 1988)). It was above all cybernetics instead of philosophy, so it seemed, that determined the fundamental formation of the modern world. This is why Heidegger held cybernetics as the moment that marked the coming to fruition of metaphysics in the form of all-encompassing technology. In cybernetics, he maintained, everything becomes actualized as an object of rational calculation. As the science of self-regulation, it was based on the instrumentalization and operationalization of communication to the extent that every central ontological relation was dispensed with (see Michel Haar, *The Song of the Earth: Heidegger and the Grounds of the History of Being*, trans. Reginald Lilly (Bloomington: Indiana University Press, 1993), 82–3). Thus, as is well known, Heidegger held that presence becomes finally completely reduced to objectness in technological domination: metaphysics has reached its end in the total technological organization of the world.

7 John Jordan, *Machine-Age Ideology: Social Engineering and American Liberalism, 1911–1939* (Chapel Hill: University of North Carolina Press, 1994), 8.

tools that ultimately embedded them in practices in science and management.[8] According to Thomas Hughes, a historian of technology, and other commentators, it was the emphasis on science and method that linked the postwar spread of the systems approach to the spread of the earlier scientific management movement, with equally revolutionary results.[9] In fact, the formal techniques developed during and after World War II, the use of which was common to all the systems sciences, also provided a response to some of the problems of Taylorism. Although claiming to be able to overcome these problems, these techniques departed from Taylorism only in their methods, not in their premises, remaining thus inside the conceptual domain of bureaucratic management.[10] Yet, not unlike Taylorism, they developed into an influential conceptual vision that immensely shaped contemporary sentiments and practices regarding communication.

These forms of knowledge, in turn, developed hand in hand with information and communication technologies that made it possible to fashion practical applications from systems-theoretical thinking. If the conceptual models of communication can be said to have their origin in the need for control, it was the coalescing environments of communications technology, computerized information processing, and their fields and methods of application, coupled with the developing disciplines of systems thinking, that marked the culmination of the control perspective from around the middle of the century. During this period, the emergence and establishment of theoretical discourses based to a great extent on these technologies allowed diverse questions and traditions to be bound together into a unified theory of control.

As for how communication became redefined as both a means of governance strategies and a subject for research, the years after World War II were crucial. At that time, a new communicative whole began to take shape as a result of an intertwining of several theoretical discourses, technological inventions, institutional arrangements, and political ends and practices. Two theoretical contributions of this period sharing the same view of communication are of foremost importance. The first can be identified as centering on the work of Norbert Wiener, whereas the second is based on the ideas of Claude E. Shannon. Taken together, these two figures laid the most important part of the theoretical foundation for a new communicative framework and specified the conceptual principles that enabled subsequent schemes for automated systems as a managerial response to

8 As Ludwig von Bertalanffy, who first gave scientific form to the systems thinking, put it: "The 19th and first half of the 20th conceived the *world a chaos*. [. . .] Now we are looking for another basic outlook on the world — *the world as organization*" (Lugwig von Bertalanffy, *General System Theory: Foundations Development Applications* (New York: Braziller, 1968), 187–8). Systems sciences sought to develop a general theory of control and self-regulation, whether in the context of a human, a machine, or a society. Other major figures in systems sciences apart from Norbert Wiener included the neurophysiologist Warren McCulloch and Jay Forrester, the founder of system dynamics. At around the same time von Bertalanffy organized the Society for General Systems Research, founded in 1954, which several mathematicians, biologists and economists joined later.

9 Thomas Hughes, *Rescuing Prometheus* (New York: Vintage Books, 1998), 8, 9.

10 Stephen Waring, *Taylorism Transformed: Scientific Management Theory since 1945* (Chapel Hill: University of North Carolina Press, 1991), 20.

the shortcomings of Taylorism to be formulated. In what follows, the language, perspective, and theoretical emphases of cybernetics and information theory will be discussed from the point of view of the thought about communication. After this, some general remarks will be made concerning the recast concept of communicativity in the new form of knowledge.

Cybernetic Constructions

One of the most central cornerstones in the postwar approach to communication, synthesizing much of the previous work done during the war on self-regulating machines, was a mathematical theory of control mechanisms by Wiener for which he coined the term *cybernetics* in 1947.[11] As a general theory of self-regulation, it created a language that, together with information theory, allowed for the integration of disciplines and themes traditionally held as separate from one another. Through this integrating language, a set of diverse questions, ranging from animal goal-seeking behavior to the operation of antiaircraft weaponry, and a whole range of disciplines such as statistical mechanics, communication engineering, the theory of control mechanisms, and biology could be brought together and dealt with within the same perspective, made coherent by the theme of "communication."[12] Thus the cybernetic discourse made it possible to compare disparate chains of questioning, fields of institutional practices, and physical machinery and technical solutions as commensurable elements of a single conceptual framework. It is hard to overrate the significance of this approach to the self-formation of a wide range of academic disciplines during a period of over three decades. As regards thinking about communication, its importance was absolutely fundamental.

Although the significance of communication from the control perspective was evident by the turn of the century, it still lacked scientific exactness and a stable definition. At that time, the commodity to be transported by a telephone system eluded rigorous description. However, a significant breakthrough was made in communications engineering when signals were separated from the electricity that carried them, and it was to this that cybernetics became connected as it gave communication a scientific definition.[13] In defining the science of communication as "the study of messages and their transmission," Wiener provided communication with a scientific core in the form of a message. Now "message" — which he defined as an "array of measurable quantities distributed in time" — constituted the elementary unit of communication, made up of the new statistical substance he called information.[14] Once the message was refined into the core substance of communication, the historical significance of communication systems seemed to

11 Norbert Wiener, *Cybernetics, or, Control and Communication in the Animal and the Machine* (New York: MIT Press, 1961 [1948]).

12 Heims, *The Cybernetics Group*, 22.

13 Flo Conway and Jim Siegelman, *Dark Hero of the Information Age: In Search of Norbert Wiener, the Father of Cybernetics* (New York: Basic Books, 2005), 65. In this respect, cybernetics depended on the previous work conducted with specification and standardization of the signals transmitted over the telephone network (see David Mindell, *Between Human and Machine: Feedback, Control, and Computing before Cybernetics* (Baltimore: Johns Hopkins University Press, 2002), 105–37).

14 Norbert Wiener, *The Extrapolation, Interpolation, and Smoothing of Stationary Time Series*, Report

decrease,[15] preparing the way for communication to be conceived of as a system in itself.

While building up a science of communication, Wiener was well aware of the central position of communication and particularly of the institutional role of it for the modern age as the "cement which binds [society's] fabric together."[16] In seeking to construct a unifying basis for the interpretation of this "cement," he interwove "communication" with "control" — concepts that appeared in the subtitle of his *Cybernetics*. In fact, these concepts were two sides of the same universal process that constituted the object of cybernetics.[17] This was the age when the "system" that has conditioned modern communication came increasingly to be explained within the frame of reference set by automated rather than purely mechanical control structures. "If the seventeenth and early eighteenth centuries are the age of clocks," Wiener proposed, "and the later eighteenth and the nineteenth centuries constitute the age of steam engines, the present time is the age of communication and control."[18] On the basis of the interconnection between these two notions through computer technology, a whole new landscape for thinking about *communis* was opened up.

Wiener combined these concepts under circumstances in which their former meanings had changed. He redefined their respective contents in a way that took into account recent improvements in technological development, particularly in computer technology, and laid the ground for thinking about communication as an automated process. The notion of communication as something with not only a systemic but also a self-organizing nature can be seen as an effort to accommodate communication to new computerized conditions. In this way, cybernetics bridged the ages of the telephone — which was a product of the industrial organization — and of the computer — which became a model for the new concept of organization. At the same time, interest in questions relating to energy and energy consumption, which largely determined the direction of analysis in the technocracy movement, had to give way to issues of communication and information. Hence, modern control theory, coupled with computer technology, seemed to synthesize the routines and practices of a given system — what efficiency engineers, culminating in the efforts of Frederick Taylor, have tried to do since

370, February 1, 1942. Records of NDRC, Sec 2, Div D. Washington, DC: National Archives & Records Service (1942), 2. See Conway and Siegelman, *Dark Hero*, 117, 127, 156, 164.

15 "Fundamentally, the social sciences are the study of the means of communication between man and man or, more generally, in a community of any sort of being. The unifying idea of these divers [*sic*] disciplines is the *message*, and not any special apparatus acting on messages," stated Wiener ("Time, Communication, and the Nervous System," *Annals* 50 (1948): 197–219).

16 Norbert Wiener, *The Human Use of Human Beings: Cybernetics and Society* (New York: Da Capo Press, 1954), 27. In the same book, Wiener argued that "society can only be understood through a study of the messages and the communication facilities which belong to it; and that in the future development of these messages and communication facilities, messages between man and machines, between machines and man, and between machine and machine, are destined to play an ever-increasing part" (Wiener, *Human Use*, 16).

17 See Conway and Siegelman, *Dark Hero*, 116–17, 183.

18 Wiener, *Cybernetics*, 39.

the beginning of the twentieth century.[19] As a result, generalized control began to shift from bureaucratic organizations to computer technology, as James Beniger has pointed out.[20] If Taylorism mechanized production, then Wiener wished to automate Taylorism.

What is absolutely central in Wiener is that his project was to formulate the logical form of exactly the same operational model facilitated by the telecommunication systems from the telegraph to the telephone and implemented in a number of organizations and industries over the previous decades. For him, this model became a general form for any system, in that every system is necessarily a communication system. Thus,

> the physical functioning of the living individual and the operation of some of the newer communication machines are precisely parallel [. . .]. Both of them have sensory receptors as one stage in their cycle of operation: that is, in both of them there exists a special apparatus for collecting information from the outer world at low energy levels, and for making it available in the operation of the individual or of the machine. [. . .] In both of them, their *performed* action on the outer world [. . .] is reported back to the central regulatory apparatus.[21]

The functional division of a system, analyzed with the help of biomechanical comparisons, was thus reorganized in the light of the principle of self-regulation. Whereas still in Taylor it was divided between decision-making and action, in which the manager was responsible for the first part and the worker for the second, self-regulation split the first of these over again. Now the decision-making element turned into two structurally distinct components concentrating on perception, on one hand, and control, on the other. So, whereas with mechanical systems the functional whole was organized on the basis of an interplay between two independent elements, the self-regulative system required a functional distinction among perception, decision-making, and action. This distinction could be achieved by three separate yet interrelated elements comprising perceptors (nerve endings), controls, and effectors, respectively. Control was thus relegated to self-regulatory structures and became immanent to its functioning.[22]

Issues such as self-regulation and control naturally have long premodern histories, but it was not until World War II that they were subjected to consistent critical review in terms of a scientific approach and regarded as a source of general metaphors. The background to interest in the theory of control and communication

19 Larry Hirschhorn, *Beyond Mechanization: Work and Technology in a Postindustrial Age* (Cambridge: MIT Press, 1984), 52.

20 James Beniger, *The Control Revolution: Technological and Economic Origins of the Information Society* (Cambridge, MA: Harvard University Press, 1986), 6. See also James Beniger, "Conceptualizing Information Technology as Organization, and Vice Versa," in *Organizations and Communication Technology*, ed. Janet Fulk and Charles Steinfield (Newbury Park: Sage, 1990).

21 Wiener, *Human Use*, 26–7.

22 Charles Dechert, "The Development of Cybernetics," in *The Social Impact of Cybernetics*, ed. Charles Dechert (University of Notre Dame, 1966), 15, 17–18.

was provided by the work done during the war on tracking radar and antiaircraft guns, which brought to the field a large number of mathematicians and physicists. At that time, hand in hand with the formulation of cybernetic theories by mathematicians, computer engineers, and neurobiologists, the formal parallels between machines dealing with information and organisms — especially the brain as part of the nervous system — formed the central object of analysis.[23] It was because of the new interest in machine–organism analogies that the nervous system was introduced again as an object of explicit theoretical analysis.

As in the era of the telegraph, it was both the structure and the function of the nervous system that was of primary interest. The structure was known as a reflex arc. In it, the sensory nerve endings, after they have detected what is happening, pass the information back to the central nervous system via the "afferent" nerve fibers, after which the nervous system passes instructions back to the muscles via the "efferent" nerve fibers.[24] What the cyberneticists were interested in was the quite automatic, or reflexive, functioning of this system, called an autonomic nervous system, to maintain constant conditions inside the body. Because the regulation process of this system corresponded to those of some machines, it seemed possible to create a general language to discuss the operations of self-maintenance in both machines and organisms. Accordingly, as a result of discussions Wiener had with Howard Aiken, one of the pioneers in computer technology, and John von Neumann, the remarkable mathematician, on the interrelationships between, for example, communication engineering and the communication and control aspects of the nervous system, the engineering and neurology aspects coalesced into essentially one because they seemed to follow the same principles.[25] Given the group's interest in control mechanisms, the central nervous system was viewed more as an automaton than as an organism: whereas Wiener treated it "like an automaton,"[26] for von Neumann it was a "large natural automation."[27] Against this perspective, it is not surprising that the thermostat was taken as the mechanical symbol of the functioning of the nervous system of the body.

During this close collaboration von Neumann and others sought to systematize the communicative functioning principle of the nervous system and devise logical models of so-called formal neuron structures that, in turn, would lay the conceptual foundation for a range of organizational structures, whether mechanical or social.[28] The nervous system and the automatic machine were considered

23 Steve Heims, "Encounter of Behavioral Sciences with New Machine-Organism Analogies in the 1940's," *Journal of the History of the Behavioral Sciences* 11, (4), (October 1975): 368, 371. For a historical summary on these themes, see also Yehuda Rav, "Perspectives on the History of the Cybernetics Movement: The Path to Current Research through the Contributions of Norbert Wiener, Warren McCulloch, and John von Neumann," *Cybernetics and Systems: An International Journal* (33), (2002).

24 Young, *Cybernetics*, 7–8.

25 Wiener, letter to Rosenblueth, January 24, 1945; cited in Peter Galison, "The Ontology of the Enemy: Norbert Wiener and the Cybernetic Vision," *Critical Inquiry* 21, (1), (1994): 248.

26 Wiener, "Time," 207.

27 John von Neumann, *The Computer and the Brain* (New Haven: Yale University Press, 2000 [1958]), 51.

28 Leonard Schwartz, "Communication and the Computing Machine," in *The Evolving Society: The Proceedings of the First Annual Conference on the Cybercultural Revolution; Cybernetics and Automation*, ed. Alice Hilton (New York: Institute for Cybercultural Research, 1966), 90.

functionally alike for they made decisions on the ground of past decisions. In both cases, information about the environment is fed back from the nerve endings/sensors over nerve fibers/closed loops to the nervous system/central control system, which directs the muscles/effectors in such a way as to maintain the desired state inside the system. Consequently, exactly as perceptors transform the energy of a stimulus into nerve impulses, the system of sensors converts environmental changes into information. Moreover, just as the brain passes instructions back to internal muscles, so the control system executes the necessary internal readjustments on the grounds of the information received. In both cases, finally, whether in an organism or in a machine, the information flows around a functionally closed system.[29] The logical relationship among perceptors, controls, and effectors, became the pivotal frame for analysis and a reservoir of more analogies for the study of communication and control mechanisms.

In this fashion, while elaborating the theoretical possibilities of the notion of the nervous system, Wiener came to the conclusion that "the closest mechanical analogue to the nervous system is the computing machine."[30] Thus the computer assumed in his thinking, but also increasingly in other sciences and popular culture cutting across cybernetic circles, a symbolical place as the generalized nervous system, a place familiar from the history of telecommunication systems. The telephone, nonetheless, retained its central position because computers constituted a proper system only through a telephone network. This popularity of the computer metaphor ran parallel with the spread of system thinking through the sciences.[31] The idea of a connection between the brain and electrical communication has, however, a history preceding the computer age, in that John Mills, for instance, spoke as early as 1934 about the telephone system as an "electrical brain."[32] Yet, it was not until the 1940s, with the development of computer scientific analyses, that this relationship could be taken as a subject for more precise theoretical investigation, and, with Wiener, it became finally a primary focus of analysis and the conceptual center for his cybernetics.

In connection with electrical communication and related discourses, it is possible to identify another source of the interest in the old biological analogy within this context. After it became known that the activity of the nervous system was based, along with some chemical interaction, on electrical discharges, the old link with the analogies provided by telecommunications rhetoric assumed an ever

29 Young, *Cybernetics*, 8–9; Robert Lilienfeld, *The Rise of Systems Theory: An Ideological Analysis* (New York: John Wiley & Sons, 1978), 65. This is also what marks the theoretical innovation of cybernetics as opposed to mechanical Taylorism, for instance. Thus, whereas an open system requires outside sources of energy and guidance, a power supply and guidance apparatus are integral in the closed system. "By this definition," writes Alice Hilton (foreword to *The Evolving Society: The Proceedings of the First Annual Conference on the Cybercultural Revolution; Cybernetics and Automation*, ed. Alice Hilton (New York: Institute for Cybercultural Research, 1966), xii), "a really automatic — *i.e.*, cybernated — system is a closed system, whereas a mechanical system is an open system."

30 Wiener, "Time," 208.

31 Paul Edwards, *The Closed World: Computers and the Politics of Discourse in Cold War America* (Cambridge: MIT Press, 1996), 192–3.

32 John Mills, *Signals and Speech in Electrical Communication* (New York: Harcourt, Brace & Co., 1934), 31–40.

more exact, scientifically justified character. The similarities between the nervous system and telecommunications made it possible to analyze both in the light of signal transmission. Thus, while it was shown in the 1930s that the intensity of the signal, caused by strong stimulation of nerve endings, was based on the time interval between successive discharges, its precise equivalent seemed to be found in telecommunications. In the latter, the same type of frequency-modulated system of signaling, in which intensities were translated into frequencies, was known as "pulse frequency modulation."[33] More generally, the joint question was about the electrical discharge — ultimately information in an electrical form — that propagates along either the nerve fiber or the coaxial cable. Yet at the same time, although a common point of departure, the electronic aspects were de-emphasized after realization of the significance of logical functions by von Neumann and others. Both nervous and communication systems are not only electrical but also, and above all, logical systems, and it was this logic that needed to be developed in mathematical form. Cybernetics, therefore, can be conceived of as a project that aimed at formalizing the nervous system by interpreting it as a mechanical principle whose logical form could be transcribed into elegant mathematical equations.

The pivotal role granted to the computer in Wiener justifies considering him as a kind of culmination in our age (insofar as computers and computer networks can be viewed as determining the age we live in) of the discontinuous sequence of development of which the first formulations were presented by William Channing in the first half of the nineteenth century in connection with the plan for the Boston telegraph system. This is to say that Wiener can be understood as a theoretician who rearticulated and reworked the organicist model to meet the requirements and promises of the computer age. At the same time the concept of the entity that unifies a community and supervises and coordinates its functions — the nervous system of a modern society — changed from being a mechanical apparatus to a digital automaton — a computer. Despite his cautiousness when trying to avoid overgeneralizations concerning the suitability of automatic control to social processes, Wiener still used the same analogy that thinkers who worked with communication systems and sought to conceptualize their functioning had used throughout the whole history of electrical communication: "[T]he computing machine furnishes us the greatest promise for an adequate central nervous system in future automatic-control machines."[34] The computer now began to dominate the conceptual terrain from which thinking about communication started out and to which it returned.

While the control mechanism of the nervous system, from the cybernetic point of view, allowed diverse disciplines and orientations to be gathered under a common concept, it was the principle of *feedback* that designated the nature of this mechanism. The thermostat became in this discourse a central conceptual node for it was the general technical model of this principle. The feedback mechanism,

33 von Neumann, *Computer*, 76–7; Young, *Cybernetics*, 4–5.
34 Wiener, "Time," 212.

which characterized the whole of cybernetic thinking to the extent that the latter can be characterized as a theory that investigates the conceptual scope and possibilities of this mechanism, has a long history that can be traced back to antiquity. In Europe, it was known in the late eighteenth century, especially in the form of the steam engine governor developed by James Watt.[35] However, the "modern" meaning of the notion of feedback is related not only to governors and self-regulating machinery but also, and more important, to the emergence of communications engineering. The notion, which subsequently strongly affected views of how a communication system functioned — especially in military defense systems — derives from the invention in 1927 of the negative feedback amplifier, made in the Bell Laboratories by Harold Stephen Black, which gave birth to the modern transmission technology of the telephone.[36] By the end of the decade, the mathematical basis of feedback structures for control devices, then called servomechanisms, was established.[37] Hence it was a question that related directly to communication — namely, to efforts to extend the telephone network across the continent and to increase the network's carrying capacity and dependability — not a theoretical problem external to it.

This principle of negative feedback, to which Harry Nyquist subsequently gave a theoretical basis, and which led to the development of the theory of servomechanisms,[38] can be seen as having been generalized by Wiener from a technical question of amplification to a general principle of a communication system. This was shown in a seminal article written by Wiener with engineer Julian Bigelow and neurobiologist Arturo Rosenblueth in 1943, in which the concept of "negative feedback" provided an essential tool for a unified description of the behavior of living organisms and machines.[39] It was in this paper, entitled "Behavior, Purpose and Teleology," in which the authors established their research topic as distinct from the mechanical systems encapsulated by scientific management, in which — unlike in the feedback systems they were set to study — "there are no signals from the goal which modify the activity of the object *in the course of the behavior.*"[40] It has been pointed out, though, that it was in the end Black's discovery that opened up a perspective from which cybernetic generalizations could evolve.[41] However, the roots of the feedback principle in telephone engineering did not go unnoticed by Wiener either.

35 Galison, "Ontology," 261–2; Hirschhorn, *Mechanization*, 32–3.

36 Black's solution was inspired by the famous engineer Charles Steinmetz, who saw the telephone company as a model for governance (Hirschhorn, *Mechanization*, 38). For an elaborate study on the history of feedback before cybernetics, see the above-mentioned Mindell, *Human and Machine*.

37 Hirschhorn, *Mechanization*, 39.

38 Robert Chapuis, *100 Years of Telephone Switching (1878–1978): Part I, Manual and Electromechanical Switching (1878–1960s)* (Amsterdam: North-Holland Publishing Co., 1982), 275, 291.

39 Arturo Rosenblueth, Norbert Wiener, and Julian Bigelow, "Behavior, Purpose and Teleology," *Philosophy of Science* 10, (1), (1943). In response to the criticism leveled against them, seven years later Rosenblueth and Wiener published a follow-up paper in which they defended their ideas; Arturo Rosenblueth and Norbert Wiener, "Purposeful and Non-Purposeful Behavior," *Philosophy of Science* 17, (4), (1950).

40 Rosenblueth, Wiener, and Bigelow, "Behavior," 19–20 (emphasis in original).

41 Hirschhorn, *Mechanization*, 39.

Wiener stuck firmly during the early phase of the cybernetic research program's formation to the thought that cybernetics as a distinct discipline or orientation was based on communications engineering. "The newer study of automata," he held, "whether in the metal or in the flesh, is a branch of communication engineering."[42] Hence cybernetics emerged and was articulated precisely from studies undertaken within the sphere of communication systems. Even all the cardinal notions such as "noise," "quantity of information," "coding technique," and so on, were taken over from communication engineering.[43] It was thus a theoretical perspective that sprang from the immanent progression and occurrence of communication itself by simultaneously reorganizing this process of development of which it was itself a part. Cybernetics, seen from this angle, represented the very self-image of communication in the electronic age.

The feedback principle, coupled with the vacuum tube, transformed the basis of machine technology from a highly constrained to a flexible one. Whereas in the former only the predetermined sequence of actions is executed, in the latter the feedback loop develops a sequence of actions that attempts to achieve the fixed outcome by making corrections in line with changing conditions.[44] So, what was unchanging in the former was invariant mechanical performance, as manifested in scientific management, while in the case of feedback-based systems, instead of a sequence of actions, the goal desired was fixed. Although such a machine responds to changes in its environment, it is, however, unimportant for the machine, as it pursues the target value through the feedback mechanism, to recognize what caused any variation in particular or its nature. Of course, this is exactly what makes this principle so efficient. When responding to its environment, this system does not recognize the cause or nature of the changes in themselves, except to the extent that they trigger continuous compensations for maintaining the prefixed outcome. The system with a feedback mechanism, hence, takes context into account as a source of information from which it can correct its own behavior. While it is not untrue to state that it is changes in the environment that control the changes in the system,[45] the environmental changes have a significance only in terms of maintenance of the system. This is why the behavior of the system is based on a prefixed norm rather than any particular context.

The concept that designated the principle against which cybernetics was directed, and hence defined cybernetics as a project, the "noise source" as it was first termed in Shannon's communication model, was later associated with the notion of *entropy*.[46] The second law of thermodynamics, according to which entropy

42 Wiener, *Cybernetics*, 42.

43 Ibid. As Flo Conway and Jim Siegelman (*Dark Hero*, 364f117) have noted, while Ralph Hartley was the first to develop the concept of noise "as an impediment in the transmission of information," (Ralph Hartley, "Transmission of Information," *Bell System Technical Journal* 7, (1928): 535–63), he did not understand communication in the statistical sense, as Wiener did.

44 Hirschhorn, *Mechanization*, 27–9. See also Gerald Flueckiger, *Control, Information, and Technological Change* (Dordrecht: Kluwer Academic Publishers, 1995), 18–21.

45 F. H. George, *Automation, Cybernetics and Society* (London: Leonard Hill, 1959), 50.

46 The German physicist Rudolf Clausius originally coined the term entropy in 1854. He defined

is steadily increasing, postulated disorganization as the inescapable fate of the world. In cybernetics, entropy appeared as a problem immanent in communication. For Wiener the major problem of communication lay in the fact that it can be distorted in transmission. The principal opponent of "normal communicative discourse," says Wiener, is the "entropic tendency of nature itself."[47] Entropy, thus, was a name for the object against which cybernetics was organized in the first place. "In control and communication," he states, "we are always fighting nature's tendency to degrade the organized and destroy the meaningful; the tendency [. . .] for entropy to increase."[48] Against this background, it was the task of cybernetics to develop a means by way of which the organized could be maintained and, if possible, enhanced. Cyberneticians saw themselves fighting against the unavoidable disintegration of order, not unlike the Clerk Maxwell demons, by devising control systems that represented islands of negative entropy in the ocean of growing disorder.[49] According to Wiener, "it is not surprising that entropy and information are negatives of one another. Information measures order and entropy measures disorder. It is indeed possible to conceive all order in terms of message."[50] This is also what brings in the theme of repetition. As it is of central importance for a communication system to develop efficient controls in order to maintain its structure and function, and, as we know from the earlier discussion above, these controls are based, in cybernetics, on the principle of feedback, the coherence of the system depends ultimately on incessant repetition of the process of feeding back. Seen from this perspective, as important as the mechanism itself is its ceaseless repetition. Wiener generalized this idea to apply to the law. According to him, the problems of law are those of "orderly and repeatable control of certain critical situations."[51] What was dear to the hearts of cyberneticians was the possibility of error-free communication and the optimal elimination of noise. While the "key to the reliable transmission of information is redundancy," noted Leonard Schwartz, "in its simplest form redundancy is repetition."[52]

What is important to notice here is that means of preserving the organized are

it as the transformation content in a thermodynamic system, referring to changes in the quantity of heat energy divided by the absolute temperature in a thermal equilibrium (see Philip Mirowski, *Machine Dreams: Economics Becomes a Cyborg Science* (Cambridge: Cambridge University Press, 2002), 46–7). As for the concept of noise, with long telephone lines it was imperative that a communication signal was not allowed to deteriorate below what was called "thermal noise" (or Johnson–Nyquist noise), first measured by John Johnson and explained theoretically by Harry Nyquist, both of Bell Laboratories, in 1928 (e.g., William Everitt, "Telecommunications: The Resource Not Depleted by Use; A Historical and Philosophical Resumé," *Proceedings of the IEEE*, 64, (9), (1976): 1296).

47 Wiener, *Human Use*, 93.

48 Ibid., 17.

49 Thus, Wiener held that "[w]e are swimming upstream against a great torrent of disorganization, which tends to reduce everything to the heat-death of equilibrium and sameness described in the second law of thermodynamics. What Maxwell, Boltzmann, and Gibbs meant by this heat death in physics has a counterpart in the ethics of Kierkegaard, who pointed out that we live in a chaotic moral universe. In this, our main obligation is to establish arbitrary enclaves of order and system" (Norbert Wiener, *I am a Mathematician"* (Cambridge: MIT Press, 1956), 324).

50 Wiener, "Time," 203.

51 Wiener, *Human Use*, 110.

52 Schwartz, "Communication," 88.

communicational in their nature. Communication represents order and meaning, against which disorder and disintegration is contrasted. This disintegration is non-communication: it is pure uncommunicativity. Because it is not within the sphere of communication, it can be conceived of only as the "other" side of communication, as its irrational part. Yet, exactly because this other side of communication is, according to Wiener, a characteristic of nature itself, communication, not unlike science, is conceived as a struggle against the disintegrating powers inherent in nature. For him, statistics became the method through which communication could be precipitated out of noise and extricated from its destructive elements.

Information as a System

In the same year as Wiener's *Cybernetics*, there also appeared a very influential communication theory by Wiener's younger colleague, communication engineer and mathematician Claude Shannon, who worked at the Bell Laboratories. It was Shannon who proposed a general model of communication and a conceptual link between communications and the computer in two journal articles that were republished a year later as *The Mathematical Theory of Communication*. This included a contribution by Warren Weaver, a mathematician and Wiener's wartime supervisor, who helped to popularize and apply the theory to the field of human communication in a separate chapter of the same book.[53] Its point of departure was in many respects indistinguishable from that of cybernetics, so the two can be viewed as sharing the same theoretical perspective toward communication, the main preoccupation being to discover the mathematical laws underlying it.[54] Shannon's theoretical project was based on the concept of "information," superseding that of "intelligence," and he approached the question of communication's measurability by using the term "bit" for the unit of information. However, although heavily indebted to Wiener's work, Shannon's concept of information diverged from Wiener's in that he held it to be the equivalent of entropy, not negative entropy as Wiener maintained.[55] Yet a theoretical basis was laid for quantitative measures of communication and of the capacity of various systems to transmit, store, and process information. Consequently, information started to be treated very much like a physical quantity.

The theory enabled accurate determination of the capacity of a given communication channel — Shannon's framing of the question was grounded in the research done on the telephone and telegraph systems[56] — with respect to the transmission of "information." Here, the communication question was approached through the

53 Claude Shannon, "The Mathematical Theory of Communication," *Bell System Technical Journal* 27 (1948) and Claude Shannon and Warren Weaver, *The Mathematical Theory of Communication* (Urbana: University of Illinois Press, 1949).

54 See J. R. Pierce, *Symbols, Signals and Noise: The Nature and Process of Communication* (London: Hutchinson of London, 1962), 42; Lilienfeld, *Systems Theory*, 50, 66. The main difference between Wiener and Shannon was that whereas for the former communication was a continuous current, for the latter it was rather an intermittent event.

55 Conway and Siegelman, *Dark Hero*, 186–7.

56 e.g., Harry Nyquist, "Certain Factors Affecting Telegraph Speed," *Bell System Technical Journal* 3 (April 1924): 324–46 and Hartley, "Transmission."

notion of a network of channels that transmit information, examples of which were not only telephone cables but also the nervous system — an enduring source of inspiration for a number of communication theorists.[57] Because the nervous system, it was shown, not only transmitted information in an electrical form but also maintained a high degree of reliability, because of the communication between elements of the systems, it became the reference system for information and control theorists, as already discussed.[58] Hence, despite the mathematical nature of the focus on communication, the biological model was never far away. On the contrary, as the most common analogy regarding communication, it was the very object of this mathematization. As in the cybernetic project more generally, it was precisely the logic of the nervous system's mechanism that produced adaptive behavior that it was sought to encapsulate in mathematical form. Communication thus received a mathematical value within the organicist framework of the nervous system.[59]

The concept of information became a general unifying thread for a whole range of different approaches to communication. In this respect, Shannon did for communication what Louis Pasteur did for hygiene, if we follow Bruno Latour's interpretation of the latter. According to Latour, the "microbe" was a tool for Pasteur to concentrate all the widely distributed and thus ineffective actions of the hygiene movement on one precise point with testable results.[60] In a similar manner, "information" served for Shannon as a means of focusing on one defined element and thus of overcoming problems that followed from attempts to take into account all possible variants and forms of communication. Information conceptualized what was universal, from the mathematical point of view, in communication, regardless of its particular type, the contents of the messages transmitted, or the persons involved. Second, it introduced a new cohesive terminology to many distinct perspectives and gathered the disparate fields of computing science, electrical engineering, physics, and molecular biology within the same conceptual terrain. Lastly, on the basis of information, Shannon rendered communication testable, which was crucial for the communication systems presupposed by modern society. By these moves, Shannon effectively standardized the field and created a uniform conceptual framework for the mathematical study of communication.

The most well-known and influential contribution of the Shannon and Weaver theory was the establishment of a conceptual model for a general communication system. It is difficult to overestimate the long-lasting hegemony it gained throughout the sciences and lines of thought that had taken communication as their object of analysis. The famous model depicts the transmission of information, or a "message," from an information source to its destination, as traveling

57 Lilienfeld, *Systems Theory*, 50. See also Jerome Wiesner, "Technology and the Telephone," in Arthur Clarke et al., *The Telephone's First Century—and Beyond* (New York: Thomas Y. Crowell Co., 1977), 29.

58 Schwartz, "Communication," 90.

59 See Steve Heims, *John von Neumann and Norbert Wiener: From Mathematics to the Technologies of Life and Death* (Cambridge: MIT Press, 1980), 302.

60 Bruno Latour, *The Pasteurization of France* (Cambridge, MA: Harvard University Press, 1988).

via a given channel from the transmitter through the receiver, with a noise source influencing the communication process.[61] On the basis of this general communication model, what was suggested as the fundamental question of communication started to be expressed in terms of a relationship between a *transmitted signal* and a *received signal*.

Although it was made clear from the outset that information or communication theory focused exclusively on the technical side of communication, the accurate transmission of electrical signals, setting aside both semantic and pragmatic questions, it is apparent that its influence was not limited to the mathematical problems of signal transmission but rather obtained its significance as a general description of communication. At the time, it was well understood among many contemporary commentators that "the theory in fact covers much more than these technical problems."[62] As a matter of fact, information theory can be seen from two different angles. On the one hand, it is an integral part of the cybernetic program in that it is the first attempt to formulate a precise mathematical definition of communication. It is this aspect of the theory that has been of the utmost importance for work done on the mathematical architecture of telecommunication systems, leading to the science of the digital computing machine. Yet, on the other hand, it demonstrated the potential for a general and consistent communication theory. In this second sense, it can be viewed as "a precise theory of communication, which may be applied to any situation where information is passed, or communication takes place."[63] Albeit having its origin in telecommunications engineering, the theory was taken to be applicable in principle to the general structure of any communicative situation: "The comprehensiveness of the [communication] theory is undoubted. It applies to all possible relationships between identifiable elements in all possible systems. It can be used for social situations of every kind."[64]

This is why the initial positioning of communication as an engineering problem and the linear model that was based on electrical communication not only set the framework for the study and construction of telephone networks but also pervaded a range of other sciences and theoretical orientations. This theory seemed to give a form to otherwise indeterminate communication processes: information, as a universal principle, "[gave] shape to the shapeless," according to Jeremy Campbell.[65] As a result, Shannon's terminology and his model were rapidly embraced and widely applied in thinking about communication in general, extending beyond the borders of information theory proper. In particular, in the

61 Shannon and Weaver, *Mathematical Theory*, 5.

62 George, *Automation*, 58.

63 Ibid., 63; see Pierce, *Symbols, Signals*, 9. The same division into limited formal applications and broader practical applications concerns cybernetics as well; see Dechert, "Development of Cybernetics," 18–19.

64 George, *Automation*, 244. In an interview in 1987, Shannon himself stated that "[i]t's possible to broadly apply the term *information theory* to all kinds of things, whether genetics or how the brain works or this and that. My original ideas were related to coding information for transmission, a much narrower thing. But some of these applications may be valid" (*Omni* magazine, August 1987).

65 Cited in Everett Rogers, *Communication Technology: The New Media in Society* (New York: Free Press, 1986), 84.

field of communication research, a linear, effects-oriented approach to communication, as adapted to the practices of a new discipline, was eagerly adopted, not least because it seemed to provide the scientific rigidity communication scholars had pursued in their attempts to professionalize the discipline.[66] There is little doubt that, for communication research, Shannon's theory was fundamental. According to Everett Rogers, the Shannon and Weaver paradigm was "the most important single turning point in the history of communication science."[67] After it had permeated the field, the theoretical perspective it encapsulated started to function as the standard view and as the basis for a growing number of revised models of the same, enduring until the 1990s.[68] It standardized the field by defining its objects, terminology, and method. It laid a whole new basis for the understanding of communication, if it was desired to take this as a scientific concept, by establishing what was its actual structure, how to determine the relationships involved, and how to interpret the results.[69]

By jettisoning what was only circumstantial and accidental, this orientation sought to identify the essential and necessary variants in the innumerable forms of diverse communication processes. It opened up a perspective within which these processes could be seen as repeating similar patterns, following the same intrinsic laws, which could now be discovered by means of two tools: the concept of information and the model for applying it. The formal approach to communication separated the relevant from the irrelevant, so that the basic communicative relationship could be explained without the need for taking into account all the possible variants and contexts of its occurrence. Yet, this logical form also produced the empirical form, insofar as researchers studied empirical communication events.

Cybernetics and information theory embraced a particular view of communication that was directed against misuse of and misunderstanding over words, while at the same time trying to set aside the power aspect inherent in the occurrence of communication. This it sought to achieve by means of a precise language, the unambiguousness of which would not allow any negotiation or discussion over the meaning of words. One can identify at least two separate, though interlinked, reasons for this. First, unequivocal linguistic relationships are regarded as necessary in order to exercise control. Although the development of a control language can be seen as both a technical and a historical question, the technical approach in fact also to a large extent sets the context for the historical question. In this sense, cybernetics (and information theory) marks a crucial point of intersection where the control dimension of language (language both as the object and the means of control) is given a technical form. Second, the demands of scientificity required cybernetics to create a description system that would allow clear and

66 Klaus Krippendorff, "The Past of Communication's Hoped-For Future," *Journal of Communication* 43, (3), (Summer 1993). Reprinted in *Defining Media Studies* (Oxford University Press, 1994), 43.

67 Rogers, *Communication Technology*, 85.

68 Krippendorff, "Hoped-For Future," 42–3.

69 Rogers, *Communication Technology*, 88, 89.

precise references to be made. An adequately objective and unequivocal system was a precondition for reaching a scientific standard, and any linguistic structure that did not meet these standards was doomed to stay outside the scientific realm proper. Here, the idea of science, of course, comes from the narrow understanding of the practices of natural sciences, the ultimate goal being a system of measurement by which all relevant objects could be viewed as commensurable and all interrelations among objects could be precisely identified, assessed, and formalized. In cybernetics, both the demands of control and scientificity were met in a language that would harmonize its internal relationships while being capable of dealing with external changes.

It is misunderstandings over words, their undecidability and simultaneous utterance that amount to noise, the principle against which information theory is directed in the first place. Michel Serres has noted that the English word *noise* was taken from an Old French word written similarly, meaning both a row and outburst of anger, on the one hand, and a quarrel or dispute, on the other. According to him, whereas English appropriated the former meaning, French retained the latter sense of the word.[70] It is the task of cybernetics, if not to totally eliminate this noise, in the sense of a discord and a disorder, then at least to regulate it according to some rational principle. This regulation can be regarded as being based on the idea of a contract. Shannon conceived communication as a contract — a regulated structure with defined relationships between the sender and the receiver. Here, one can recall the analogical relationship that there was between the telegraph as an intercommunication system and the contract principle: both the sending and the receiving stations, and ultimately the communicating persons, on the one hand, and the contracting parties on the other, had the same interest in combating the surrounding noise that could drown their own voices. It is this same need to stabilize and formalize interrelational conditions and thereby control noise that can be found at the core of Shannon's conception of communication.

Contracts define the boundaries of a common area.[71] In the dominant tradition within which communication has been viewed as an interrelationship between independent communicating parties, Shannon reintroduced the idea of a contract as a counter to noise by giving it not a technical or juridical but an exact mathematical formulation; all changes "in the transmitted signal," in his theory, were called "*noise*," the common name for all that could impair the transmission.[72] Now the legitimacy of the contractual principle was no longer dependent on a formal authority but instead was built into the very communication system, the task of which was to guarantee that the desired signals pass through the entire system. The contract consisted of the promise to ensure the best possible reception of a message sent over a given channel and thus bounded a closed circuit between the communicating parties. This set aside, to an extent, the material conditions of

70 See Michel Serres, *Le contrat naturel*, Éditions François Bourin, 1990 (*The Natural Contract*, trans. Elizabeth MacArthur and William Paulson, Ann Arbor, MI: Michigan University Press, 1995).

71 The concept of contract, as used by Serres, comes close to the "Weltbild" of the late Wittgenstein, which constitutes a framework for agreements and disagreements concerning the truth.

72 Shannon and Weaver, *Mathematical Theory*, 8.

communication, from the point of view of the parties involved, and made the communicative situation look transparent and technically mundane — perhaps one reason why telephony, for instance, has not so far been sufficiently studied from the constitutive point of view. Yet this state of affairs inevitably links communication inherently to law understood as a regulated set of positions and relationships by which multiplicity and polyphony is closed and stabilized into formalized rules.

Both cybernetics and information theory sought to formalize a mechanism by means of which a system would deter disorder. The former approach was based on gun-control research in antiaircraft weaponry, understood as a problem of information theory. Each shell fired was regarded as a message, and information about its accuracy was used to direct the next shot.[73] If for Wiener the criterion for the success of "messages" was their hit rate, for Shannon it was the exact reproduction of the signals transmitted. Communication, for Shannon, was fundamentally about reproducing at one point a message sent at another point.[74] Therefore, in both cases, a strictly determined relationship between a starting point and an end point, in terms of a message to be sent, was the foundation for the cohesion of the system. Anything that could compromise this relationship would undermine the identity of the whole system.

Hence, if disintegrated and inefficient practices represented for Taylor a source of disturbance and disunity that were to be eliminated by a scientific approach, the enemy for cybernetics and information theory was "entropy" and "noise," respectively.[75] From different yet interlinked angles they defined what was essentially a mutual problem. These notions pointed to the two principal subject areas against which mechanisms of exclusion were set in motion, as practiced in cybernetics and information theory. Thus, it was the task of cybernetics to exclude disorganization and loss of meaning. In fact, control systems were seen as islands of negative entropy that were running against the main flow of the tide of disintegration, thus preserving order and meaning.[76] As for information theory, it excluded, first, distortion and, second, interruption — factors that could prevent accurate reproduction of a signal at the receiving end. The aim of both was to engender and maintain a delimited area of order in order to allow a coherent communicational environment within a permanent state of unfixedness and decay. Entropy and noise, in consequence, did not remain outside the area of order; rather, they constituted the ever-present tendency against which order was established and

73 Rogers, *Communication Technology*, 94.

74 Shannon and Weaver, *Mathematical Theory*, 31. See also Flueckiger, *Control, Information*, 101–2.

75 However, the concept of *entropy* was also important for information theory, being defined by Shannon as a measure of variability. On the other hand, the notion of *noise* was also regarded as a cardinal one in cybernetics. Yet, the approach to noise was somehow different between these orientations in that whereas the former held that it was bad in itself, in the latter it gained a more positive status as a prerequisite for a self-correcting system. Mirowski (*Machine Dreams*, 76n48) suggested that the reason why noise was presented as bad in information theory was a throwback to the military origins of what he calls "the cyborg sciences."

76 George, *Automation*, 56.

constantly maintained. Nevertheless, they defined the other side of order and organization and, at the same time, the source of their identity.

To the extent that cybernetics and information theory brought about a concept of communication that was closely tied to the objective of maintaining order, they can be said as being a privileged example of subordination of what cannot be presented by any system of representation, and what has been variously termed by different theorists a multiplicity, becoming, or difference, for instance. This multiplicity, polyphony, or incommunicable — that is, politics — is here a derivative term, before which comes, first, complete transmission, the perfect correspondence between the transmitted and received message. This identity of the message as maintained over the line represented the pure form of communication. Second, before the multiplicity but after complete transmission, comes a message that has been exposed to disturbance, being incompletely received, distorted, and garbled. This second type of communication represents what should be avoided, a kind of failure, against which the identity, truthfulness, and authenticity of the message is mirrored and determined. Thus, the enemy proper, in this scheme, is the noise, which corresponds to what is unpredictable, indeterminable, undecidable, and which hides behind every meaningful message as a threatening and chaotic principle. It is this very noise (politics) against which information theory deployed its strongest weapons and defenses. It made a cause out of defending order, anticipation, and clarity against ambiguity and unpredictability.

Yet one of Shannon's fundamental ideas was precisely that noise is not external to communication, but is rather always inherent in it. It was his aim to calculate the values that would show how information could be best preserved during its dispatch from sender to receiver. If information was garbled because of noise, one should either amplify or repair the flow of bits. Noise, then, was an integral part of the communication process, and information theory a tool to preserve and defend communication from itself; that is, from the disintegrating tendencies intrinsic to it. By thus separating communication from its noise, mathematics enabled communication as a pure form. The "other" side of communication in information theory, therefore, was nothing other than communication's limit as intelligible communication.

Similarly, in cybernetics, external variation was both a source of error that must be eliminated and a factor that continuously checks and corrects the coherence of the system. Hence it represents, on the one hand, a hostile element the influence of which is to be abolished in order for equilibrium to be restored, but on the other hand it is itself a prerequisite for this self-correcting system. This is because without it the system would not have any method to control its own behavior. Without this change in its environment the system would not have any means to monitor and control its own functioning, and it is the monitoring and control that makes what the cybernetic system is about.

Given this, it would be too simplifying to assert that if in Taylorism the most efficient mode of action established the standard for expending time and removed the possibility of exercising one's discretion over work, then, in the same way, Shannon excluded the noise from the communication system. This is because just as rest is a precondition of work and cannot be regarded as independent of the activity involved, noise is a precondition of communication. Similarly all the individual and disordered constituents that were the focus of attempts at systematization and uniformity proved resources necessary to the system's functionality. Nevertheless, the major problem of communication remained the fact that messages can be garbled or distorted in transmission.

In the final analysis, the idea of a pure mediation or complete, faultless transmission is nothing other than the Enlightenment's conception of a world entirely permeated and illuminated by truth. Yet, as shown earlier, because it was understood that distortion could not be totally eliminated from communication, the issue become more complicated. Insofar as this distortion was caused by human activity, and not just by "nature itself," it belonged to the realm of power, or that of politics. An indisputable mechanism was to exclude the effects of multiplicity on action, in the form of law. "[Law] is the process of adjusting the 'couplings,'" stated Wiener, "connecting the behavior of different individuals in such a way that what we call justice may be accomplished, and disputes may be avoided, or at least adjudicated."[77] Thus it was law that defined any optimal relationship *in* communication. But even the law was not insulated from the effects of distortion. What annoyed Wiener was the politics, or "game," as he termed it, inscribed into the practice of law, the fact that the conflicting parties seek to "jam" the messages of the opposing side, as the situation was seen from his viewpoint. Politics, alongside legal practices, was for him a game that only jammed the movement and reception of intrinsically unequivocal messages. If human life, however, is not conceivable, in principle, without politics in some form or another, then apparently distortion is a necessary part of it. As Wiener and Shannon were provenly aware of this, their project must essentially be regarded as an attempt to find a model of equilibrium between action and the necessary conditions that are always part of the action itself.

In its attempt to avoid ambiguity and vagueness, the cybernetic project (including information theory) was forced to view natural languages as too imprecise to allow the exact values and relationships required. The need for technical terminology rather than ordinary usage of words and precise syntax rather than natural linguistic grammar resulted in manufactured or artificial languages, especially in the mathematical representation system. The measures or values introduced by mathematization were not external, but always immanent to communication. The mathematical relationships it produced were communication's own relationships. Yet, the history of communication as an object and means of governance is not bound by the mathematical domain. Shannon's contribution, seen from this perspective, rather created a whole new standard for the study of communication that it was sought to generalize, as Weaver's original step in this direction proves, to human communication while still trying to retain scientific rigor.

By and through Shannon's work, communication, understood as information, was irrevocably connected to the realm of mathematics. It is because of the

77 Wiener, *Human Use*, 105.

immanent logic of communication expressible as mathematical formulas that coding, transmitting, and decoding of signal sequences — crucial for designing and understanding self-regulating systems — becomes possible. Since this connection was made, it has been almost impossible for systematic thinking on communication to escape from the sway of this theory. Of course, communication as science cannot be said to be attributable to Shannon's contribution alone, for it has its own history of knowledge irreducible to information theory. However, Shannon can be regarded as having perfected the project of quantification, or rationalization, of communication, the roots of which can be traced back at least to Ward but which only got its modern *sense* through the increasing need for and influence of formal analysis. Though the systematic formalization of communication can be said to have its inception at least in the requirements imposed by early telegraph systems, it was not until mathematical communication theory arrived that communication became a quantitatively determinable subject in a scientific sense.

Practices of Autocommunication

From the perspective of signal transmission, a system based on the feedback principle is circular, and circles are precisely what determined the conceptual domain of cybernetics. Recursive circular processes characterized not only more or less confined systems but also, in principle, the social realm, viewed as a system, which makes this perspective a general world-relation. "The social system," we recall, was, for Wiener, "bound together by a system of communication, and [. . .] it has a dynamics in which circular processes of a feedback nature play an important part."[78] In a circular information structure like this, action is manifested in terms of messages transmitted, processed, and fed back. It is no accident, then, that the dominant conception of what communication is has predominantly been message-driven since cybernetics and information theory.[79]

Now, taking cybernetics as a more or less continuous project, the ever-refining model of a circle can be regarded as not just a theoretical contribution but also, more fundamentally, as the next major break, rupture, or displacement, after the scientific management, in the history of communication and control. If communication, in scientific management, was based on sequences, here it is grounded in circles. This theoretical turn is reminiscent of one of a more practical nature effected by William Channing in relation to telegraph lines. As already shown, whereas telegraph lines were used to connect different locations distant from each other by means of direct lines, Channing, when planning the municipal telegraph system of Boston, twisted the lines together into a complex and multilevel circular structure. This step, although clearly taking place in a different age and context, implies a similar aim to that at play in the move from Taylor's sequences to Wiener's loops. The circular structuration, intrinsic to communication systems, is given a more exact theoretical formulation in Wiener, and if the explanation and understanding of this structuration were based, in Channing's initiative, mainly

78 Wiener, *Cybernetics*, 24.
79 Krippendorff, "Hoped-For Future," 42.

but not entirely on practical reasons, in Wiener's case it was approached through an already institutionalized field of communications engineering.

The theoretical legacy of cybernetics, taken broadly, was crystallized in the view that central to communication is its circular pattern, generated by its repeatable control, not its specific content. This idea was manifest in all self-regulating systems in the form of feedback loops. How communication is organized according to these loops depends on control systems, which is why the focus was on the theoretical investigation of the mechanics of control. As Shannon's communication model spread and became common, communication appeared more and more as a patterned formula in which the mechanisms to code and decode the flows of information came to the fore.

To perceive cybernetics as a study of self-governance under the circumstances of constant change required it to be understood within the realm of control. For it was against change, unforeseen and uncontrollable transformation, that formal approaches to communication as a tool of control, the history of which has been traced here, were constructed. Within this line of development, form has always been pitted against content, and method against goal. This was evident, for example, in the way the Turing test distinguished the content of a communication process from its form, thereby laying the foundation for the principle that any precisely specified problem can be solved by a computer.[80] Whereas in scientific management the logic of action resided in the formal organization of the functional whole, in cybernetics it was built into the behavior of the system itself; yet, in both cases, whether a hierarchy or a self-regulatory system, a preset norm, albeit of a different nature, differentiated what was acceptable from what was unacceptable. In both cases, in addition, the "we" of the action was determined prior to the action itself.

What connects Wiener and Taylor is the fact that they approached their respective systems presupposing a certain methodological identity or consistency, whether in the form of "negative feedback" or "the one best way," that gathered chains of action together under one system. Taylor preceded Wiener in the respect that he had already in use a logically closed system that tried to anticipate and incorporate all possible states in a given context. Of course, the nature of these thinkers' respective systems differed, but what they shared was an idea of a closed, harmonious, and uniform structure that could maintain its identity over time. The essential ideal for both cybernetics and Taylorism was to design a system that would eliminate any need or purpose for negotiations over a course of action. Instead of imposing a foreign principle forcefully on a system, they sought the immanent laws inside it and tried to incorporate them in a standardized terminology. The measures, models, and units derived ultimately from communication itself, not outside it.

This allows us to view cybernetics and information theory as interlinked efforts to formulate a theoretical foundation for the process of self-organization, understood as a general principle of communication. In this respect, it follows

80 Edwards, *Closed World*, 159.

previous attempts to propose the telephone system as the model of governance for a continuously changing system. An issue here is the old theme of rational self-maintenance of a community under constant change — that is, a conscious self-governance in which the means of control are exposed to a process of incessant reorganization. The structure of communication, as it were, was here the entity that controlled its own operation. Yet it was not communication as such that was of central interest, but rather the *mechanisms* of communication that would determine the limits of a given system.

If Taylor tied work and communication to law, which was what controlled the forms of these two interrelated processes, then Wiener and Shannon, for their part, mathematized this law. Their law was no less real than Taylor's one as its formal structure is nothing other than the expression of the law in a framework that was a precondition of modern communication. The unchanging and original principles behind all communicative action were interpreted as being from beginning to end of a mathematical nature, which was what allowed automated control of communication. It was not until after the 1970s, it can be said, that communication was freed from subjection to a founding law. Then the structural conditions of communication began to break away from a horizon of closure and gravitate toward openness, multiplicity, and flexibility.

In the case of a mechanical structure, the line between a system and its outside, or a norm and a disorder, or even the field of truth and that of corruption and distortion, was drawn by a formally determined set of interrelationships and positions. The decision concerning the structure of internal relationships was made prior to, but not independently of, the action proper. It maintained unvarying links among its components while excluding all redundant or unwanted actions, thereby producing a system with invariant functioning.[81] Thus, its point of view was *exterior* to communication. In essence, it was an articulation of communication prior to its happening, from a perspective insulated from communication.

When the principle of self-regulation made this unchanging system, unable to deal with change, responsive to its environment, the concept of change was defined as part of the purview of the system itself. Environmental change, or the multiplicity that the system must cope with, does not modify in any way in systems based on the feedback principle, the prefixed goal or the internal state desired. Rather, it is the system, then, that constantly maintains its unity and identity against the environment. Yet the point of view here, in contrast to that of a mechanical system, is *interior* to communication. The articulation of communication takes place within the sphere of communication itself, in the course of and during realization of the interrelationships between the system and its exterior, by means of continuous compensations. This is what is meant by the fact that control mechanisms were not seen as uncommunicative with respect to the information they regulate, but rather as having themselves communicative nature, as pointed out by Karl Deutsch, Wiener's colleague at MIT (Massachusetts Institute of Technology),

81 Hirschhorn, *Mechanization*, 16.

following Wiener on this.[82] Although a distinction has been made between a signal and information, the signal being the operation and information the control of a communication system,[83] it is clear that both information and signals constitute a system in which communication is both its functioning and control of this functioning. From this point on, communication became a self-modifying process.[84]

If scientific management, too, operated from the basis that control of action was mainly communicative, here control was incorporated into the communicative structure of the action itself as a self-correcting principle. Within an information system, communication, through various physical, social, and conceptual solutions, set the norms for its own performance. It can be argued that, even in the first cybernetic machines, both performance and the rules controlling this performance were subject to adjustment in the course of the system's functioning.[85] So in cybernetics, communication and its conditions were mutually defined — within certain predetermined limits that indicated the "purpose" of the system — through the process of communication becoming articulated. It was communication and its mechanism, now, that defined what was fundamental, not the rules or hierarchies that were supposed to tie an action to a system. However, this latter aspect remained necessary, in one form or another, to the formal approach to communication.

Yet, a change in thinking about communication occurred when communication as a continuous, self-constitutive process began to take shape. This was made evident by a conceptual shift that took place within the cybernetic project away from emphasis on equilibrium and homeostasis and toward the concept of feedback and reflexivity.[86] Although it was necessary to continue thinking about communication through a conceptual whole, this whole became more porous and reflective in relation to its environment. In effect, this introduced a more profound change in the perception of communication.

The systems approach tied communication to a perspective in which it no longer featured in terms of relationships between independent subjects but in which these communicative relationships created the subject positions involved.[87]

82 Karl Deutsch, *The Nerves of Government* (Glencoe, IL: Free Press, 1966), 77.

83 Pierre de Latil, *Thinking by Machine: A Study of Cybernetics*, trans. Y. M. Golla (London: Scientific Book Guild, 1956), 31.

84 See Karl Deutsch, "Mechanism, Organism, and Society: Some Models in Natural and Social Science," *Philosophy of Science* 18, (3), (1951): 240; Karl Deutsch, "Mechanism, Teleology, and Mind," *Philosophy and Phenomenological Research* 12, (2), (1951): 193. In his definition of feedback, Deutsch states that by the term he means a communications network which "includes the results of its own action in the new information by which it modifies its subsequent behavior" ("Mechanism, Organism," 246; "Mechanism, Teleology," 197).

85 Galison, "Ontology," 258.

86 See N. Katherine Hayles, *How We Became Posthuman: Virtual Bodies in Cybernetics, Literature, and Informatics* (Chicago: University of Chicago Press, 1999). The term "homeostasis" was coined by the neurophysiologist Walter Cannon, a friend of Wiener's father (see Conway and Siegelman, *Dark Hero*, 15; Walter Cannon, *The Wisdom of the Body* (New York: W. W. Norton & Co., 1939 [1932]).

87 Of course, although with different motives and goals, the collectivist turn in the progressive era could be thought of, in a certain sense, as a predecessor of this shift. As already mentioned, this turn implied an ontological change in the conception of the individual from that of an independent entity to being defined through a broader social framework (see Bentley, *Process*, 204, 206; Schwendinger and Schwendinger,

Communication was no longer, contrary to conventional wording, a bond between distinct subjects. It is true that the status of communication in relation to community still tended to be expressed in this way, which explains Wiener's references to communication as the "cement" that binds society's fabric together.[88] Here, communication was regarded as a kind of addition — albeit an important one — to a preexisting social whole, the actions of which it then adjusted and bound together. While this view was presented in various forms in the twentieth century, coincidental with the rise of systems thinking if not earlier, communication itself became gradually the central focus of attention.

Now, as pointed out, these perspectives are not mutually exclusive, since it was the dimension of control that unified the systems approach to communication. Rather, control now found for itself an immanent basis in the intrinsic nature of communication: control became internalized into the self-execution of communication. This was linked to a more general process in the course of which communicative wholes were transformed into information systems. They were transformed into these systems not just in the sense that information was more important to their functioning than energy, but rather in the sense that their functions were none other than the functions of information — the interactive gathering, storing, processing, and transmission of information.

This is what allows us to regard this era as one of the most important periods of change during which a new way of thinking about communication came into being. Here, communication was no longer about interrelationships between preexisting subjects, as had been largely the case in discourses and practices relating to the telegraph and to the early telephone. Neither was it *exclusively* about the original whole that tied the subjects closely to the system, a system that enabled them to be articulated as subjects, as had been the case in the age of the telephone and scientific management. The new conception, though, insofar as it was linked with systems thinking, could not shake off the idea that the social domain was a determinable, pre-given totality.

From the point of view of cybernetics, insofar as communication is defined primarily as recurrent transmission and control of information, then the "automechanisms" of a system — in other words, the techniques and practices that maintain its identity — are based on the transmission of messages and on their serving as information for other messages, thus directing messages round in endless circles. Importantly, these mechanisms constitute a system, and are not merely subsequent additions to it. Therefore the system can exist only because of these mechanisms; it is nothing that could be disentangled from them in principle.

Sociologists of the Chair, 123, 193–4; Lustig, *Corporate Liberalism*, 110–29). Now the systems approach could be considered, up to a point, a continuation of this tradition in having redefined it in terms of communication as a logical structure. Yet, even if systems theory could be seen as having supported the collectivist approach in purely communicative terms, the latter remained fairly short-lived and limited: social-choice theory, based on the premise of individual rationality, gained ground in social thought after World War II in opposition to theories supporting communism or collectivism (see Kenneth Arrow's epoch-making book *Social Choice and Individual Values* (New York: Wiley, 1951).

88 Wiener, *Human Use*, 27.

This brings us to what is new in the assemblage of communication based on systems thinking. Although clearly remaining stuck in the perspective of an original conceptual whole, because of the communicational nature of controls, thought about communication was, however, directed from the beginning from the point of view of the "between" that communication is. Communication was not approached solely from the viewpoint of communicating entities but rather through communication itself.[89] Yet this relation to thinking about communication in itself that was opened up by cybernetics remained essentially just a possibility, the potential of which remained to be utilized at a later time. This question was brought up again in "second-order cybernetics," in which observation of an action and the action itself formed a circular system that presupposed no stable ontological basis.[90] This orientation was manifest specifically in the so-called "new" biology that sought to dissolve the categorical opposition between "organism" and "environment," and instead stressed their "co-evolution."[91] The argument goes that the environment cannot be separated from what organisms are and that their relationship is rather organized on a progression of feedback loops, the products of which are meaningful only in connection with this circular process. In other words, the process and its products are intelligible only with reference to one another. Thus these "autopoietic" processes set and define their own boundaries in a chain of mutual referential relations,[92] with the effect that the "system" and its "environment" can no longer be categorically separated.

Following this shift in perspective the conception of communication changed for good. The technocratic dimensions of cybernetics, particularly in the form of the pursuit of unidirectional control, were called into question. Effectively, there was a change in focus from causality to autonomy and from action and reaction to interaction.[93] Of course, it can be pointed out with good reason that the profound change in the theory of knowledge that seemed to occur here was already implicit

89 However, as already noted, the progressive era was influential in its efforts in this direction. Since this period the "modern" history of communication has abounded with individual theorists who have pondered upon communication as such, with John Dewey providing perhaps the most significant attempt to formulate it as constitutive of a community. Here, however, I am thinking about the first established and widespread practice that brought out the self-constitutive nature of communication.

90 e.g., Bernard Scott, "Second-Order Cybernetics: An Historical Introduction," *Kybernetes* 33, (9/10), (2004): 1365–78. The next wave of the cybernetic program can be identified with Heinz von Foerster's second-order cybernetics, according to which people construct their models out of the systems they work upon all the time. From then on began a broader application of cybernetics and systems thinking to a number of distinct fields, including psychology and the study of social systems. This second wave reached its mature phase with Humberto Maturana and Francisco Varela's book *Autopoiesis and Cognition: The Realization of the Living* (Dordrecht: D. Reidel, 1980), which articulated the view that organisms, in addition to being self-organizing and self-regulating, are also autopoietic, or self-making.

91 See Maturana and Varela, *Autopoiesis*.

92 Keith Pearson, *Germinal Life: The Difference and Repetition of Deleuze* (New York: Routledge, 1999), 146–7, 169. See also John Mingers, *Self-Producing Systems: Implications and Applications of Autopoiesis* (New York: Plenum Press, 1995). According to Niklas Luhmann, "[c]ommunication is only possible as an autopoietic system" (Niklas Luhmann, "How Can the Mind Participate in Communication?," in *Materialities of Communication*, ed. Hans Gumbrecht and K. Ludwig Pfeiffer, trans. William Whobrey (Stanford: Stanford University Press, 1994), 379).

93 Ranulph Glanville, "A Cybernetic Musing: In the Animal and the Machine," *Cybernetics and Human Knowing* 4, (4), (1997).

in Shannon's information theory, if not earlier.[94] This is because the receiver is not capable of distinguishing between noise and the signal, in his view, independently of a theory that makes this possible. It is on the basis of the theory alone that the distinction can be made; the theory determines what is the signal and what is noise, as well as their mutual relationships, within a particular communication channel. This theoretical shift, in which communication was not regarded any longer as independent of its context, became in any case the explicit starting point for second-order cybernetics, from which it received a detailed theoretical treatment. Yet, what it shared with classical cybernetics was the presupposition concerning the organizational closure of these autopoietic systems.[95]

The cybernetic system was essentially a model of equilibrium, as it represented a mechanism that enabled a dynamic system to be in harmony with its environment while continuously adjusting itself to the changes. Although remaining within the realm of closed systems, it positioned the mechanisms of enclosure and exclusion as immanent to the operation of communication. Once control became immanent to communication, however, interest in studies concerning how control was manifested empirically gained increasing grounding in theoretical discourses on communication, undermining the significance of a general theory of control. Second, a general theory of the cybernetic type could no longer provide a satisfactory framework for thinking about communication, because communication started to be thought of without a single constituting whole. This is what gave communication an independent existence, as it were.[96] Finally, research on artificial intelligence started to gain ground in the late 1950s, and supplanted cybernetics as the natural theoretical framework for questions of communication and control.

Yet the model of cybernetics, it should be kept in mind, was not that of a self-sufficient system in a mechanical sense. Whereas for a mechanical system changes represented a potential menace, for a feedback-based system they were essential. This was because this system was based on importing differences, not on their exclusion as was in the case of mechanical structures. Consequently, not only do the intrinsic processes draw a distinction between the system and its exteriority, thus creating a difference, but also they take this difference as a foundation of their functioning. Environmental variations, then, are as an origin of differences, without which the system could not be functional. For this reason, the principle of equilibrium can be said to have given an impetus to the development in which the noise and multiplicity are no longer viewed as interference to the system but instead as a necessity.

It is not incorrect, however, to argue that what has been at stake in the history of communication, seen from the viewpoint taken here, has been formalization of interrelationships between the system and its outside. During this history, these

94 Constructive theories of different varieties have found a systematic conceptualization in philosophy at least since the "critiques" of Immanuel Kant. What concerns us here, however, is the way mutual articulation of an "action" and its "outside" has been realized within and in connection to the thought about communication as an object and means of governance.

95 Pearson, *Germinal Life*, 149, 169.

96 Flueckiger, *Control, Information*, 21.

relationships have been defined and organized on the basis of a method intrinsic to, and thus characteristic of, the system. The method is thus what drew a demarcation line within a set of elements between those that were conducive to a given goal and those that were not, in other words between communication and noise.

Today, however, noise is not seen exclusively as an enemy of order, it is rather its basis or source. Richard House, following Henri Atlan, brought up the idea of self-organization from noise,[97] noting that the Shannonian theory of noise applies to systems in which the received message comprises all the information the receiver has. According to Shannon, the full reproducibility of a message set the theoretical reference point against which all the rest, in other words noise, came to be assessed. Atlan, on the other hand, showed that noise plays a necessary part in communication because it is only against noise that the signal can be distinguished in the first place.[98] Self-organization from noise pertains to a situation in which the discrepancy between the received and the transmitted signal is taken as the foundation of information about the system. If both the transmitted and the received messages are known, then it is possible to draw conclusions, on this basis, about the functioning of the system as a whole. If the transferring of a message is a process in which the changes in it allow measurement of the system as a whole, the relation to noise also changes decidedly because now it becomes the basis of information concerning the workings of the system.[99] This characterizes the more general theoretical turn taking place: the formal qualities of a given object can no longer be distinguished from the perspective from which it is observed. In other words, the outside is twisted into the inside, or noise into communication, as its integral part. Thus, one cannot get rid of noise, because through it communication gets its expression. Noise is also something that cannot be shared among the communicating parties: it escapes what is common, rational, and the same. It is the necessary but "irrational" part of communication, always a threat yet absolutely indispensable.

Self-Regulating Systems

After World War II, a close tie was forged between communication and politics. Communication became an even weightier political question, and politics became articulated more and more in terms of communication. This manifested itself in different forms and contexts. On the institutional level, the relationship appeared, perhaps in its most visible form, as one between science and government. The study of communication broadly understood — that is, the diverse disciplines from communication engineering to computer science as well as from psychology to linguistics — were all harnessed to serve national interests during the war, and retained the close connection afterward.[100] But the relationship between politics and communication in the postwar period also manifested itself in terms

97 Richard House, "The Encyclopedia Complex: Contemporary Narratives of Information," *SubStance* 92, (2), (2000): 27.

98 Henri Atlan, "On a Formal Definition of Organization," *Journal of Theoretical Biology* 45, (1974).

99 House, "Encyclopedia Complex," 27. See also Mirowski, *Machine Dreams*, 76.

100 e.g., Heims, *von Neumann*, 179; Heims, *The Cybernetics Group*, 3. A canonical justification for

of general questions concerning the politics of control. The theories formulated by Wiener and Shannon constituted a unifying conceptual terrain within which questions of control, whether as practices of business management or national discourses on political governance, could be posed in terms of communication. As in the case of Taylorism, the new language was embraced in distinct projects and systems, although now their scope had grown in size and complexity. And as in Taylorism, what was significant in these projects was not their specificity but, on the contrary, their becoming part of a common perspective in which a number of other projects were planned, compared, and accomplished. Communication in the form of new technologies, theories, and practices became a key field of systematic development during the Cold War era, and it was to a large extent cybernetics and information theory that laid down the general outlines from which this development work took its models.

Although certainly varying in their particular aims and methods, different projects, from local to nationwide in scope, reflected models that were essentially the same in their inherent characteristics. It was a model based on the deep-rooted metaphor of the nervous system that had remained recognizable while undergoing reinterpretation by cybernetic theorists. Thus the elaborated nervous system constituted, in the last analysis, the form in which different political interests, technical arrangements, and theoretical discourses were thought of and, taken together, articulated as a consistent whole. If it is true that the nervous system was precisely what allowed communication to be thought about in itself, it was no doubt also the whole in which various social questions and technological imperatives of the postwar period encountered.[101]

The discourse on generalized control found expression in and through diverse systems being developed throughout the country, but, at the same time, construction of these systems was propelled to a great extent by this same vision of control. With large-scale computerization and the formation of specific fields and techniques out of and alongside the theoretical foundation laid down by Wiener and Shannon, the perspective of technological control acquired a political meaningfulness that seemed able to overcome the previous difficulties linked with Taylorism. The computer, as a "tool for managing the mass society,"[102] coupled with new mathematical instruments, became a new social icon because it could keep up with the increasing number and pace of processes resulting from technological development. Now, the cybernetic technology of governance that established an immanent mechanism between communication and control seemed to be applicable, for the first time, to managerial problems in general, not only to those in

state patronage of scientific research in this period was provided by the influential book by Vannevar Bush: *Science: The Endless Frontier* (Washington, DC: Government Printing Office, 1945).

101 This was true, in particular, of efforts to grasp communication as such, in its totality. Thus it was only natural for Harold Lasswell, for instance, in his attempt to portray communication "as a whole in relation to the entire social process," to place his account within this terrain of biological analogies (Harold Lasswell, "The Structure and Function of Communication in Society," in *The Communication of Ideas*, ed. Lyman Bryson (New York: Institute for Religious and Social Studies, 1948), 203–43.

102 Daniel Bell, "The Social Framework of the Information Society," in *The Microelectronics Revolution*, ed. Tom Forester (Oxford: Basil Blackwell, 1980), 509.

some limited expert systems, provided that the problems were operationalizable in principle.

This depended on the possibility of completing the project of mathematization of the social realm in its essential characteristics, as it were. Attempts to find mathematical laws governing communication, following and extending the program initiated by Wiener and Shannon, found expression in a number of systems-scientific disciplines and discursive practices. The general objective, based on communication theory, was to "produce a predictive theory of humans and machines in group activity, and in certain industrial situations."[103]

Society certainly did not change into an automatic machine based on formal rules during this era of automation. Yet the extent of how the idea of this kind of machine shaped social thinking and various industrial practices cannot be overestimated. Cybernetics transformed, according to Flo Conway and Jim Siegelman, "the day-to-day labors of workers in every industry."[104] Yet many of the concepts dealt with in this study on control and communication have been influential in that they have contributed to the formalization not only of communication but also of the resources required to grasp what is essential in society. My concern has not been, however, to show the specific routes and mechanisms through which the distinct concepts were generalized and spread throughout the society. It is doubtful if that is feasible at all. In other words, I have not analyzed the forms through which administrative relationships have been generalized and turned into immanent communicative relationships.

Instead, what I have done here is to examine the discursive forms (metaphors, theories, models) inherently related to the development of modern communication systems. These discourses have been considered as part of the historical line along which communication has been taken as an object and means of control and along which different relationships have been forged with the limits of communication. This does not mean, however, that I have been operating solely within the discursive sphere either. Certainly, these discursive practices and forms of knowledge do have concrete effects on the social realm, being in many ways constitutive of it. The underlying idea here has been that the social in its different aspects is indistinguishable from communication. This is why the processes of development have been approached as assemblages consisting of certain discursive forms, institutional frameworks, subject positions, national politics, cultural beliefs and habits, physical structures, and organizatory methods, all of which get their identities as part of the interrelationships, and especially the *ethos*, that give coherence to the given constellation.

In what follows, I will address the practices of automation — that is, the theoretical discourses, institutional routines, and technical systems that together allowed the society to be conceived, in principle, as a self-regulating system with management of its key sectors and problem-areas to be automated one at a time. After this, I will look at the way the communicative constellation changed and

103 George, *Automation*, 247.
104 Conway and Siegelman, *Dark Hero*, x.

the new form of communication this gave rise to. Let me start, however, from the local and national information and communications systems that allowed contemporaries to cherish the idea of the birth of a rational and predictable social environment within the framework of a nervous system.

Constructing Information Systems

The language of cybernetics gained popularity as the new communication and computer technologies became common.[105] It was through this language that computer power and the politics of management were combined in the same perspective. For Wiener it was evident that "the importance of information and communication as mechanisms of organization proceeds beyond the individual into the community."[106] Yet he was reluctant to generalize the methods of cybernetics to the social domain as such, as many social processes seemed to remain beyond the scope of operationalization. Although Wiener admitted that it was possible to formulate a mathematical description only for some of the social and communicational processes, for him such a description was the main criterion for their scientificity. If the social was not mathematizable in itself, the conceptual clarification of its formal properties as well as the uncovering of its logical relationships was in any case the only way to scientific understanding of society.[107] Although Wiener did not believe in the operationalizability of social sciences, he certainly would have wished it to be possible. For only if this had been the case would the social sciences and historiology have achieved, in his eyes, the standards of objectivity required for a proper science. Until that moment, according to Wiener, "there is much which we must leave, *whether we like it or not*, to the un-'scientific,' narrative method of the professional historian."[108] Regardless of some of his reservations, he thought, at the end of the day, that cybernetic principles were more applicable in social sciences the more they assumed quantitative methods.[109]

It was left to Wiener's followers to test the applicability and explanatory capacity of cybernetics to social problems, especially from the management perspective. In the 1950s, cybernetics and the sciences influenced by it almost invariably included social organizations and even entire nations as a relevant subject.[110] This is why

105 Heims, *The Cybernetics Group*, 271.

106 Wiener, *Cybernetics*, 18.

107 Norbert Wiener, *God and Golem, Inc.: A Comment on Certain Points where Cybernetics Impinges on Religion* (Cambridge: MIT Press, 1964), 88. See also Reinhold Martin, "The Organizational Complex: Cybernetics, Space, Discourse," *Assemblage* 37, (1998): 103–4.

108 Wiener, *Cybernetics*, 164; emphasis added.

109 Wiener, *God and Golem*, 87–90.

110 David Easton, for instance, applied cybernetics to political theory (David Easton, *The Political System: An Inquiry into the State of Political Science* (New York: Knopf, 1953); David Easton, *A Framework for Political Analysis* (Englewood Cliffs: Prentice-Hall, 1965), Karl Deutsch applied it to the art and science of government (Deutsch, *Nerves of Government*), and the economist Kenneth Boulding used cybernetic principles for solving other social problems (Boulding, *Organizational Revolution*). In a similar vein, Marshall McLuhan, the famous communications theorist, used Wiener's ideas concerning the role of messages in society in elaborating his views on electronic media (Marshall McLuhan, *Understanding Media: The Extensions of Man* (New York: McGraw Hill, 1964); Marshall McLuhan and Quentin Fiore, *The Medium is the Massage: An Inventory of Effects* (New York: Bantam, 1967); see Conway and Siegelman, *Dark Hero*, 277; Martin, "Organizational Complex," 109). However, the more clearly information was thought about,

we have to turn to these individual organizations and projects as links or points of oscillation between specific interests, practices, and problems, on the one hand, and the instances of realization and development of what was a general conceptual perspective of governance, on the other. Thus, in the following I will investigate in more detail the main directions through which self-regulating systems came to occupy an ever more central position in the representations of society.

The Cold War era was marked by the design and implementation of ever more complex national information systems, especially in the automation of air defense. In his study on the speech about national defense systems, and construction of them during the Cold War era, Paul Edwards showed the way in which the history of the computing machine was linked with the defense needs of the United States through the concepts and discursive practices of cybernetics, artificial intelligence, and cognitive psychology.[111] In all these fields a central aim was to outline the feasibility of constructing a system in which all circumstances of a given area of life, in this case the concern over a nuclear attack, would be taken into account and programmed as possible variables enabling the system to respond automatically and largely independently of human interference. In this way, discourses and related practices maintained the idea of a kind of transparent and self-reflective world wherein all the major characteristics of potential states of affairs and their mutual relationships were prearranged so that the system would be able to retain automatic control and dependable functioning.

The first large-scale, computerized command, control, and communications system, *SAGE* (Semi-Automatic Ground Environment), which became fully operational in 1961, divided the United States into eight distinct sectors, each controlled by a real-time information processing combat center. Not unlike the models based on functional similarities with formal neuron structures, it formed a huge integrated whole with simultaneous information flows on many hierarchical levels organized into one logical structure. Its two main functions were to gather information and to direct defense. Radar systems would detect all aircraft and feed their positions over telephone lines to be received, processed, and evaluated by the combat centers responsible for generating an overall view of the situation at hand. The direction centers, for their part, each connected to a radar installation, would produce surveillance information to assign and direct the defense systems.[112]

The important point here is that SAGE provided not only a technical basis for the development of military systems but also a discursive environment for speech about such systems, reinforcing the need to create a centralized defense arrangement. In this way, computer technology and the related discourses that

according to the classical cybernetic tradition, as a concept from which the semantic dimension was excluded, the more the problems concerning the mathematization of the social domain were emphasized. Therefore, the application of cybernetic principles to social problems was a trend that also gave rise to critical opposing reactions among many contemporary social thinkers. For instance, the sociologist Donald Macrae argued that the social universe did not provide a suitable population for the mathematical expression of the relations with which cybernetics was most concerned (Donald Macrae, "Cybernetics and Social Science," *British Journal of Sociology*, 2, (2), (1951): 148).

111 Edwards, *Closed World*.

112 Ibid., 75–111; Hughes, *Rescuing Prometheus*, 52–4; Mirowski, *Machine Dreams*, 350–1.

represented the world as a logically closed system were mutually articulated in an inherent relationship with each other in which each shaped the development of the other. It was SAGE that laid a foundation for all future military as well as industrial computer projects working toward centrally coordinated, real-time systems. In this way, it solidified a certain systems-scientific language and ideology.[113] Most notable among the computer projects, whether operational or only as prototypes in the late 1960s, were the American Stock Exchange's computerized quotation service, *Amquote*, the American Airline's automated reservation system, *SABRE*, and the American Bankers Association's still unnamed real-time system.[114] For all these systems, the overall aim and method of managing all the processes would be essentially the same, involving the need to keep up with constant change in a complex environment by means of an integrated computer system.

The operational position of centrally directed communication systems grew in importance with the increase in power of electronic digital computers. Even though he himself was careful not to directly support centralization of power,[115] Wiener was still inclined to think about the undeniable benefits of a centrally controlled social system. At any rate, not only the increasing presence of systems based on computers[116] but also the presupposition, central to Wiener's thinking, of the role of a control mechanism to maintain the stability of the system, influenced in the shaping of the frame of analysis in which ideas relating to and inspired by cybernetic research came to be expressed in ever more full-blooded formulations.

By the early 1950s, it was possible to describe the communicative diagram of the whole of society in cybernetic terms. Thus, according to Kenneth Boulding, professor of economics at the University of Michigan, "the barest essential of any organization [. . .] is a control mechanism consisting of a system of communication and transformation."[117] His views can be seen as an expression of how in cybernetics a system predicated on the organismic analogy was formulated into a principle of an automatically functioning whole. A communication system, according to Boulding, consisted, first, of *receptors of information* (thermometers, sense organs, or salesmen were included in the examples given) that collected information about the environment and transmitted it over a *communications system* (nerves, speech, letters) to an *executive* (furnace control, central nervous system, manager) whose business it was to take the received information and to transform it into *instructions*, which were finally communicated to *effectors* (furnaces, muscles, workmen) who, following the instructions, adapted their behavior to the environment or even changed it.[118] For Wiener's followers, communication as an automatic system was generalized into a generic model, linked

113 Lebow, *Information Highways*, 174; Edwards, *Closed World*, 110.

114 Robert Macbride, *The Automated State: Computer Systems as a New Force in Society* (Philadelphia: Chilton Book Co., 1967), 9–12.

115 Heims, *The Cybernetics Group*, 180.

116 There were only "computers" at the time we are talking of. The epithet "mainframe" was not coined until minicomputers started to be built at the very end of the 1960s and early 1970s.

117 Boulding, *Organizational Revolution*, 68.

118 Ibid.

up with basically the same biological concepts as were operational from the early telecommunication systems onward.

Clearly, SAGE was a concrete manifestation of this functional structure. It embodied conceptual elements from two directions: first from the ideas that were inherited from the old biomechanical metaphors, and second from those that were elaborated within cybernetics and information theory, leading to the systems sciences. These sources, of course, were not independent of each other, as I have tried to show in this study. On the contrary, the nervous system was exactly what the latter theoretical tradition aimed at formalizing, interpreting it as a mechanical principle of control. Yet both the use of biological metaphors and the tradition based on the contributions by Wiener and Shannon do have distinct historical origins, albeit sharing, at the same time, also many points of contact. In SAGE these ingredients seemed to have been given a solid physical manifestation. It was not an existing structure, the nature of which the ideas relating to biological metaphors and theories based on them would then attempt to explain. Quite the reverse, it was a purpose-built defense system that was predicated on this very discursive tradition.

SAGE largely defined the functions a national computer system would need. Those were identified by a contemporary commentator as follows: "a set of 'sensors,' or remote terminals, located wherever information from local sources is fed into the system," "a central 'exchange,' where information from local sources is assembled, sorted, and stored and dispatched to those who need to use it," "a central 'control unit' which generates programs and specifications for the use of information," and "a set of 'outlets' — again, remote terminals — located throughout the country and particularly throughout Government departments, to which orders, instructions, and commands are dispatched."[119] In a sense, then, SAGE was the first uniform nationwide realization of the ideas being developed in the discursive fields that aimed to find and codify the immanent principles of automated communication systems. It was, in other words and in the eyes of many contemporaries, the automated nervous system of the nation.

It was during this period that the most grandiose schemes for rationally managing diverse sets of transactions and processes by using real-time computer systems were proposed, constructed, and put into use. If computer systems resembled telephone systems in that both allowed multiple access, the former also made a break with the latter. Whereas the telephone system only enabled connections to be made, the computer system monitored the content of each message.[120] This was something different from the telephone's linear way of operation. Now information generated at many different sources could be used as a source for new information: it could be fed, processed, and used as a basis for subsequent, perhaps completely automated, messages to be sent to remotely located devices. What is more, not only could a potentially immense system be controlled by computer but also it could control itself within certain pre-given limits. Computers, then, helped control

119 Macbride, *Automated State*, 104.
120 Ibid., 13.

become internalized into the system itself by directing message traffic throughout the system and by generating new instructions out of the data processed: it was the system's instrument of self-governance. The biomechanical theme thus assumed a new, computerized identity. As Gerard Piel, publisher of the *Scientific American*, put it in the 1960s:

> The new development in our technology is the replacement of the human nervous system by automatic controls and by the computer that ultimately integrates the functions of the automatic control units at each point in the production process. The human muscle began to be disengaged from the productive process at least a hundred years ago. Now the human nervous system is being disengaged.[121]

The theme of an automated system of governance and control presented in the framework of a nervous system was mainly articulated in three distinct yet interconnected discourses. Let me now take a brief look at each of these fields. The first deals with "automation" and methods of refining a self-correcting principle, applicable to management of the social realm; the second is the vision of a national scheme emerging out of the grid of local systems; and the third is a perspective characterized by the end of ideology, thanks to the comprehensive national system with its political methods of self-regulation.

1) "Automation" was one of the most influential topics that marked social discourse until the 1960s, when it became known as "technological change."[122] It constituted the prevailing conceptual framework through which technology appeared as a key political question. The notion of automation, coined in 1947 by Delmar S. Harder, the vice president of the Ford Motor Company, and brought to broader public attention by John Diebold in 1952,[123] provided a compact way of conceptualizing what was characteristic of a time fueled by enthusiasm regarding the potential and challenges of computers. In contrast to Harder's original meaning of the notion — "handling parts between successive production operations" — Diebold used the word to mean "both automatic operation and the process of making things automatic."[124] It signified the substitution of mere mechanization by something qualitatively different; that is, by a process with self-regulatory characteristics.

121 Cited in Robert Theobald, "Cybernetics and the Problems of Social Reorganization," in *The Social Impact of Cybernetics*, ed. Charles Dechert (Notre Dame: University of Notre Dame Press, 1966), 45–6.

122 Macbride, *Automated State*, 54f.

123 E. M. Grabbe, "The Language of Automation," in *Automation in Business and Industry*, ed. E. M. Grabbe (New York: John Wiley & Sons, 1957), 20; John Diebold, *Automation: The Advent of the Automatic Factory* (New York, 1952). However, Margaret Marshall pointed out that the term was used as early as in the seventeenth century by Simon Patrick, a minor Cambridge Platonist. By "automation" he meant an all-inclusive mechanism that simulated the smaller mechanism of the clock on the scale of the universe (Margaret Marshall, "'Automation' Today and in 1662," *American Speech* 32, (2), (1957): 149–51).

124 Cited in Marshall, "Automation," 149. Three years later, in 1955, the term automation was defined by Walter Buckingham as signifying "any continuous and integrated operation of a rationalized production system which uses electronic or other equipment to regulate and co-ordinate the quality and quantity of production" (US Congress, Subcommittee on Economic Stabilization of the Joint Committee on the Economic

One of the first public statements of the problems posed by automation and the computer revolution, entitled "The Triple Revolution: An Appraisal of the Major U.S. Crises and Proposals for Action," was published not until 1964.[125] It spoke fervently in favor of managing the economic destiny of the nation, especially in connection with the immense power of the "developing cybernated system."[126] Within the same year, President Lyndon B. Johnson established the National Commission on Technology, Automation, and Economic Progress, which issued its first report in 1966.[127] It was through these and similar accounts that, on the level of national politics, the concept of automation started to become solidified. At the same time, computers were seen not only as one of the most important single sources of sociotechnical change, but also as a means of controlling the consequences of this change and the attendant instability. That is, computers as a means to control their own effects.[128]

This view had already been introduced, though not elaborated further, in the proceedings of the National Advanced-Technology Management Conference held in 1962, in which the issue was presented in terms of communication. So, whereas "in spite of improvements in communication media [. . .] managers of government and industrial organizations find communication an increasing problem," on the one hand, "technical breakthroughs in data processing equipment have provided an opportunity for development of integrated systems of information flow," on the other.[129] Thus, once a new technology called "information technology"[130] had become common, the computer assumed the same strategic role as the telephone earlier, as both the cause and the cure for increasing instability. Yet as a result of experience gained from the telephone, attempts were made to use the computer to harmonize its own adverse consequences far more purposefully than was the case with the earlier system. By appearing to be the major social control

Report, *Hearings on Automation and Technological Change*, 84th Congress, 1st Session, 1955, Statement by Walter S. Buckingham Jr., 32). While the distinction between mechanization and automation gradually became clear, it was Daniel Bell in his elegant definition who put forward the core idea that automation is related specifically to computerized control: "Automation is a process which substitutes programmed machine-controlled operations for human manipulations. It is the fruit, so to speak, of cybernetics and computers" (Daniel Bell, preface to *The Age of Automation*, by Sir Leon Bagrit (New York: Mentor, 1965), xvii).

125 Macbride, *Automated State*, 46–7.

126 Cited in ibid., 48. "The aim throughout," declared the Memorandum, "will be the conscious and rational direction of economic life by planning institutions under democratic control" (cited in ibid., 54).

127 Ibid., 54–5.

128 Theobald, "Cybernetics," 47. Of course, as is known, this assumption of the computer as a tool rather than a medium has been effectively criticized, especially in recent critical media theory. A perceptive example of this type of criticism is Friedrich Kittler, "Thinking Colours and/or Machines," *Theory, Culture & Society* 23, (7/8), (2006): 39–50.

129 Fremont Kast and James Rosenzweig, eds., *Science, Technology, and Management* (New York: McGraw-Hill Book Co., 1963), 306.

130 One of the first uses of the term "information technology" is to be found in Harold Leavitt and Thomas Whisler, "Management in the 1980s," *Harvard Business Review* 36, (6), (1958): 41–8. In this paper, the authors defined the term as including three related parts: 1) techniques for processing large amounts of information rapidly, 2) the application of statistical and mathematical methods to decision-making problems, and 3) the simulation of higher-order thinking through computer programs; Leavitt and Whisler, "Management," 41.

devices, computers seemed to form the "essential stepping stones to the Great Society."[131]

The issue was considered mainly in terms of the effects of automation on unemployment, which constituted generally the framework within which the computer was treated as a national question at that time.[132] One of the first statements suggesting a computerized nationwide service to ease the threat of growing unemployment was in a testimony given by Isaac Auerbach to the Senate Subcommittee on Employment and Manpower in 1963.[133] The question was crystallized in a concrete proposal for an "early-warning employment-prediction system," a computerized nationwide service for matching people to jobs. This President Johnson's Great Society project was included in the Automation Commission's report, as mentioned, where it was depicted as a complex, hierarchical system to coordinate manpower allocation: "With local centers feeding into regional centers information relevant at that level, and these in turn feeding into a nationwide job and manpower bank, the service could provide detailed information on the manpower requirements of job vacancies and the personal characteristics of job seekers."[134] There was no need, however, to limit this or subsequently proposed systems to considerations of employment alone. These systems in industry and government, in fact, gradually moved society closer to one that was increasingly suffused with discussion on sophisticated automation and advanced self-regulation. The "early-warning employment-prediction system" was clearly planned to be the first civilian SAGE, aimed at automating the workings of a crucial sector of the society.

This embodied and made concrete the idea of how "computer-caused problems are met with computer-based solutions," as Robert Macbride, a commentator writing about the social effects of computerization, put it in 1967.[135] Computer technology facilitated new quantitative approaches and planning schemes for forecasting and coordinating societal processes: mathematical models of these processes began to be used at the federal level along with new computer and management information systems and the contribution of physical and social scientists

131 Macbride, *Automated State*, 90.

132 e.g., Dirk de Wit, *The Shaping of Automation: A Historical Analysis of the Interaction between Technology and Organization, 1950–1985* (Hilversum: Verloren, 1994), 13, 83. For discussion in the 1950s and 1960s, see, for example, R. L. Meier, "Automatism in the American Economy," *Journal of Business* 29, (1), (1956): 14–27; Reuben Slesinger, "The Pace of Automation: An American View," *Journal of Industrial Economics* 6, (3), (1958): 241–61; Robert Lekachman, *The Age of Keynes* (New York: Random House, 1966), 226–45; Vladimir Stoikov, "Increasing Structural Unemployment Re-Examined," *Industrial and Labor Relations Review* 19, (3), (1966): 368–76. For a discussion about the relationship between automation and managerial control, see Thomas Whisler and George Shultz, "Automation and the Management Process," *Annals of the American Academy of Political and Social Science* 340, (1), (1962): 81–9.

133 Auerbach stated, "it is my belief that it is possible to develop an early-warning employment-prediction system based on 'gaming' techniques." The final report of the Subcommittee also considered the implications of this kind of system, formulating the issue as a question: "Can the very devices which are playing an important role in present changes in manpower requirements contribute to the solution of manpower problems by adding to the accessibility and analysis of manpower data?" (cited in Macbride, *Automated State*, 64, 65).

134 Cited in Macbride, *Automated State*, 63.

135 Ibid., 75.

who had been integrated into American public policy.[136] The ultimate vision was of a social structure that could maintain and control itself, to a point, with the help of computers. By their availability, so it was thought, it would be possible "to spot trends long before they would otherwise be visible, to carry out the necessary discussion and to develop policies before the need for action develops."[137] This was not a depiction of a state to be reached sometime in the future but of a mechanism that, practically speaking, was already there. All the necessary technical devices for social self-regulation were already in place, it was believed, and now it was only a question of how they would be connected together and assembled into a whole. The local systems already developed constituted the crucial precondition for this structure.

2) The evolution of local government computer systems was, after its initial disorganized phase, rapid and comprehensive. One of the early governmental systems was the New York Police Department's "CORRAL," which went into operation in 1965. It was a computer surveillance and detection system aimed at identifying traffic offenders and stolen cars on the streets. Once the computer recognized the license number that was fed into it by a police team stationed on a busy highway as one of a "wanted" vehicle, it reported the match back to the team who, then, sent the information to a second police team further up the road, which stopped the vehicle. This was a starting point for the planning of a "total system," in which the computer would not be just an operational tool but, rather, become a means for integrating the functions of the whole New York City Police Department, according to Captain Adam D'Alessandro of the Planning Board of that department.[138] Yet even this "total system" marked only the beginning of a remarkably large-scale project wherein all local information systems would merge into one another and form an immense network for the federal government. As it seemed in the late 1960s, "the future utilization of a vast centralized computer system encompassing in some respects virtually all of the ferment of the local government systems, and, to a lesser extent, most big commercial systems, is as probable as the use of its decentralized alternative."[139]

What would best serve as a model for this national computer system? Although SAGE naturally set the pattern for future real-time systems, it seemed to be too purpose built to suit as a model for a system with a necessarily more unconnected and heterogeneous structure. This is because the national system would be assembled out of existing arrangements and in response to a more complex situation, whereas SAGE had been built as an individual project in response to a well-determined task. At this point, the Bell System was revived as a paradigm again. What was of interest here was how a balance was achieved between the functions of existing individual companies and the management of the network

136 Wilson, *National Planning*, 56–7.
137 Theobald, "Cybernetics," 47.
138 Macbride, *Automated State*, 93–4, 95–6.
139 Ibid., 102–3.

that integrated these units into a whole. In the Bell System, the innumerable subsidiaries throughout the country operated with considerable autonomy, while AT&T, as the owner of the companies, took care of technical aspects and ensured compatibility among all these subsystems.[140] This solution seemed to point the way for the emerging national system, too.

It became evident in the 1960s that computer systems were not limited to a few specialized applications, but were bound to permeate every part of society. The line of development from a computer solution for a specific problem to a whole range of practices, discourses, and institutions enabled by a computer implied the establishment of a new framework for thought and action in the midst of the communicative relationships largely based on linear telephonic assemblages. The computer, thus, like the telegraph and the telephone previously, was not merely an auxiliary device but, rather, a precondition for modern society and for the communicative practices and forms of self-description characteristic of it. It was recognized that one had to acknowledge the reality and constitutive power of the computer, especially in the form of large-scale digital computer systems, and "the inevitability of their evolving into a unified national system affecting every important socioeconomic area of our lives."[141]

In the computer's becoming commonplace, then, a relationship to the ontology of communication opened up: out of the basically mundane and repetitive routines of gathering, storing, classifying, and comparing information evolved a network of ideas and practices that was seen to be essential to society. It was within this framework that the growth and expansion of computerized relationships and processes could be seen as part of an integrated system. But, on the other hand, the spread of computerization formed the meaning-content of the integrative perspective. It could not be distinguished from this perspective because it constituted exactly what this general system was. This framework, furthermore, did not originate in the technological sphere alone, as if consisting only of rapid computerization and automation, seen as a process that reduces to specific devices and machinery. Rather, it was the result of a mutual articulation of technological systems and discursive practices. This led to the traditional concept of politics coming to an end, or at least so it appeared in systems thinking.

3) Within the later cybernetic tradition, the "political sphere" as such seemed to be amenable to systematization and formalization. Every social and political institution was given a specific function within an automated system. The whole social system began to appear, in this discourse, as a kind of equilibrium in which inputs were transformed into outputs through a certain preset hierarchy of functions: politics seemed to become a mechanism that was regulated by chains of receptors, controls, and effectors, inputs and outputs.[142] What is noteworthy is

140 Ibid., 105.

141 Ibid., 173.

142 Thus, "a constitution is a program defining the nature (activities) and interrelations of the formal loci of political power. The outputs of the political system are enforceable laws defining the interrelations of persons and groups within the society. Demands on the political system are communicated by petition,

that relationships in the realm of politics were understood as being entirely predetermined in their functional characteristics. Political institutions were reduced to a single pre-given function, which they could not redefine and beyond which they could not move. Basically, then, it was a closed system in which, if the desired state can be agreed on (which is, of course, the endless task of politics), would maintain this state by comparing information about the overall behavior with the given norms and by making any corrections necessary. What is more, the whole process would be carried out without disputes, differences of opinion, and endless negotiations.

The belief that a neutral technique — mathematics in particular — could separate administration from politics became a hallmark of the "end of ideology" in the late 1950s.[143] Political decisions were evaluated in terms of technical efficacy, with the scientific approach setting the standards for success. The underlying presupposition, no doubt, derived from the vision of a perfect self-enclosed machine. Within a total system of this sort, overall performance could be stabilized in spite of outside disturbances, so it was thought. Here, the influence of these disturbances, constituting a potential risk to the functioning of the system as a single determined whole, has been actively neutralized by the feedback principle, which maintains fixed order in changing circumstances. This has been the aim of numerous influential attempts at social planning and system engineering since Taylorism at least. Throughout history, projects to develop a perfect social design, predicated on a presumed set of invariant principles that, "if represented in community rules, would create a self-regulating social life, free of conflict and change," have occurred.[144] Especially during what has come to be called modernity, these projects, accompanied by and articulated in connection with a scientific authority, have assumed an immense role and brought about profound effects on how the social system has been thought of.

This is perhaps the culmination of the history of communication as an object and means of control, as the explicit undoing of politics, in the sense of seeking principles of control that would not admit several interpretations, turns here to politics itself as an institution. Despite the original reservations concerning the applicability of cybernetics to social problems, this view already figured strongly in Wiener's writings and highlighted society as such as the focus of thought. Wiener considered cybernetics precisely as a general social theory, not as a specialized discipline: the concepts of communication and control revealed the common principles that can be found in humans, machines, and whole communities, too. Here cybernetics followed the discourses that aimed to conceptualize the history

by representatives of organized groups, by publicists, and other means including elections. Legislative decisions are made in the form of laws and resolutions. The executive puts the laws into effect and the judiciary serves a control function by comparing specific individual actions with the law that programs such action" (Dechert, "Development of Cybernetics," 25).

143 Hanson, *Democratic Imagination*, 295. A canonical book on this theme is Daniel Bell, *The End of Ideology: On The Exhaustion of Political Ideas in the Fifties* (Cambridge, MA: Harvard University Press, 1960).

144 Hirschhorn, *Mechanization*, 57. See also Robert Boguslaw, *The New Utopians: A Study of System Design and Social Change* (Englewood Cliffs: Prentice-Hall, 1965), 2.

of modern communication into general formulations and models, which are always at the same time comments on society itself and its conditions. Although extremely complex in comparison with what Taylor could have imagined, communication as a pattern continued to be thought of in terms of not only a structure but also a perfectly rational one that, so it seemed, brought politics to an end. As it became susceptible to automation through the concept of negative feedback, communication appeared to be an action implementing in a reflex-like manner the functional unity of a system and receiving its meaningfulness from its pre-given purposes. Not until about 20 years later was a discursive genre established that emphasized that communication engenders relationships and processes that defy rationalization, and, moreover, that communication by its nature is not something that can be thought of or captured as a whole.

The idea of society as a closed circuit found cogent expression in Wiener, who considered society essentially as a communication system. He held that "society can only be understood through a study of the messages and the communication facilities which belong to it." The stance that communications are the integrating element of society, the cement that binds its fabric together,[145] was repeated in various works influenced by Wiener's ideas. If communication was the cement that made organizations and societies, as Wiener put it, and if it was inherently unequivocal, which was what the cyberneticians had to presuppose in order to pursue the laws of communication, it could in principle be codified into a consistent set of rules. These rules were exactly what the cybernetic project produced as its theoretical legacy and what bound this project to the tradition followed in this study. Thus, according to the aforementioned Karl Deutsch, for example, these rules, although far more complicated in social communication than in machine communication, nevertheless conformed to the same grammar.[146] Society, in this project, became a machine, and in contrast to the machine analogies in Taylorism, was capable of continuous self-correction. Similarly, communication in its pure form became a control function.

Managing Decision-Making

In the course of the development through which computer systems came to form a general framework for communication and control, certain discourses emerged centering on these systems as managerial tools. These discourses sprang also from changes that occurred in business organization and in the environment in which businesses operated. As organizations grew in size and complexity, resource coordination and allocation had to be rethought. The Taylorian solution based on "diagrams and maps" would no longer be adequate. More detailed and timely information was needed in order for management to keep up with complicated and rapid organizational processes.[147] As a result, the systems approach was widely introduced

145 Wiener, *Human Use*, 25.

146 Deutsch, *Nerves of Government*, 77.

147 L. D. Introna, "Management Information Systems," in *Encyclopedia of Communication and Information*, ed. J. R. Schement (New York: Macmillan Reference, USA, 2001).

into managerial practices as a response to the shortcomings of Taylorism, especially as the systems way of thinking met the needs and aims of technocratic managers and social planners. The reason for the rapid spread of the systems approach resided in its potential, as seen by the enthusiasts, to govern complex sociotechnical systems, the functioning of which could be controlled and monitored through computers. Systems sciences and planning techniques together created a view of progress according to which the rationalization of the factory accomplished by Taylor could be extended to the management function, in this way finally removing the flaws in decision-making based on subjective and parochial politicking.[148] In what follows, I will show the general line of development along which ideas based on cybernetics and information theory were instituted in the managerial practices, systems, and discourses, in the framework of what came to be called "management-by-computer."[149]

Systems sciences, operations research being one of the most important if not the foremost of these, created a conceptual basis for rethinking organizational strategies in terms of formal models. By the mid-1960s, they — together with information systems and "strategic planning" — had begun to dominate the thinking in American corporations and American government.[150] They created a language of their own that linked numerous institutions, technical arrangements, and discursive practices with the shared objective of quantitative problem management and a rather mechanical ideal of governance. Faith in an apolitical and unequivocal system of governance by mathematical means not only made systems theorists part of the tradition of control of and by communication but also it marked the redefinition of this tradition because these means internalized control as part of the structure of action. Control was now an immanent quality of the self-performance of communication. As the ideal of mechanical governance increasingly lost favor in the 1970s, the idea of control being immanent to communication remained, albeit in different forms.

Operations research — subsequently transmuted into systems analysis — formulated an academic language to translate problems to be tackled into technical issues. Its roots were, like those of cybernetics, in the work done with military defense and control systems. In 1951, Philip Morse and George Kimball synthesized methods used during the war so that they could be applied to business problems.[151] Their work laid the ground, along with the first textbook by C. West Churchman, Russell L. Ackoff, and E. Leonard Arnoff,[152] for the institutionalization of a new theoretical line of thought that converted military techniques into standardized ones for civilian use. It aimed at formulating a "scientific basis for solving problems," replacing mere "experience," "intuition," "opinion, vague

148 Hughes, *Rescuing Prometheus*, 142–3, 166; Henry Mintzberg, *The Rise and Fall of Strategic Planning* (London: FT Prentice Hall, 2000), 21, 23.

149 Macbride, *Automated State*, 7.

150 Mintzberg, *Strategic Planning*, 6. See also Mirowski, *Machine Dreams*, 165, 177–90.

151 Philip Morse and George Kimball, *Methods of Operations Research* (New York: John Wiley & Sons, 1951).

152 C. West Churchman, Russell Ackoff, and Leonard Arnoff, *Introduction to Operations Research* (New York: John Wiley & Sons, 1957).

generalities, or lore."[153] Thus, the foe of scientific problem-solving technique was not unlike that of Taylorism. Indeed, what was shared by all projects that sought to formalize managerial methods was the attempt to substitute inefficiency and disputes originating from subjective beliefs and inherited habits with a standardized, systematic approach based on "scientific" procedures. It was faith in formal techniques that took Taylorism, according to Waring, to a logical conclusion in the systems approach.[154] For this reason it has been argued that no meaningful line of demarcation can be drawn between operations research and scientific management or between scientific management and management science. Common to all was a persistent search for quantification and systems in management.[155]

Some differences, however, can be identified. Whereas the Taylorists saw a system as composed of isolated tasks and chains of operations, operations researchers viewed it in terms of interrelated mathematical quantities and relationships. The systems metaphors that permeated operations research were appropriated from information theory and cybernetics, in which systems were fundamentally communication systems and communication was information expressible in mathematical form. Moreover, the system, which for the Taylorists consisted mainly of the accomplishment of a set of tasks, appeared to operations researchers rather as a way to solve problems.[156] In short, then, although operations research was a response to similar problems to those in Taylorism, it implied an altered concept of a system. Now, the system was viewed as 1) a thoroughly interconnected pattern that 2) consisted of mathematical relationships and 3) concerned problem-solving. While retaining the whole as consistent, and thus maintaining connections to the historical figures of speech relating to communication systems, it differed not only from scientific management but also from cybernetics in a crucial way. In cybernetics, the system, although equally mathematical in its nature, remained contained within a realm of expert systems, broadly speaking. In operations research, however, the processes and relationships being optimized were no longer those of more or less limited systems but those of the social realm in itself. Therefore, they came to cover all kinds of managerial problems, provided that they were operationalizable in principle.

The growing use of operations research with its mathematical techniques for solving managerial problems intensified in the 1960s, with AT&T becoming one of its major corporate users. It was largely the faith that managerial problems could be scientifically formulated, modeled, and solved, thus eliminating intuition in decision-making for standardized routines, that encouraged the use of operations

153 Churchman, Ackoff, and Arnoff, *Operations Research*, 6. Cited in Waring, *Taylorism Transformed*, 24, 25. See also Boguslaw, *New Utopians*, 49–51.

154 Waring, *Taylorism Transformed*, 47. In Mirowski's interpretation operations research is "little more than an extension of the Taylorist movement of 'scientific management'" (Mirowski, *Machine Dreams*, 178).

155 Herbert Simon, *The Shape of Automation for Men and Management* (New York: Harper & Row, 1965), 69; Hughes, *Rescuing Prometheus*, 8. For an account of the history of operations research and its relationship to Taylorism, see M. Fortun and S. S. Schweber, "Scientists and the Legacy of World War II: The Case of Operations Research (OR)," *Social Studies of Science* 23, (4), (1993): 595–642.

156 Waring, *Taylorism Transformed*, 25–6.

research in business.[157] Concurrently with the development of computer technology, operations research was institutionalized within the corporate domain as experts hired originally on a temporary basis started to be appointed as staff specialists with a permanent assignment. The growing capacity of computers to process large amounts of data encouraged establishment of internal departments to rationalize planning for a whole organization.[158] With increased speed and improved efficiency, computerized systems processed more operations and did so more rapidly, creating new information that was utilized to solve other problems, thus continuously stimulating the whole process. In this way, computer systems themselves, while performing their operations, became a source of new information already in machine-readable form that was used as a basis for new operations. Here, the crucial point was that, because of the automated systems, information could be used about information, not just about things.[159] Thus information became autonomous to a point in that it was not only the means but also an object of operations. As a result of this development, self-organizing domains started to spring up, based on information that worked on information. And while it was on this domain that new management practices attempted to capitalize, its coming into being was largely due to systematic and intensive subjection of communication to administrative procedures along the lines followed in this study. Therefore, the domain of information, understood in this sense, was a result of historical developments in which communication has played a double role as both an object and a means of control.

The introduction of special departments for organizing and managing corporate information systems created a need to develop and systematize managerial practices in large organizations. This led to the organization of information so as to facilitate managerial decision-making. Many corporations began to actively utilize their growing capacity in information processing and to sell it to other corporate buyers. This reshaped the theoretical understanding through which information used to be conceived. In contrast to Shannonian information theory, the developers of information systems, as well as the later systems approach in general, were interested in the meaning and effect of information. This approach broadened the linear communication model of Shannon by dealing "not with one communication link at a time but with a large number of them simultaneously, not with binary communication relations between a single sender and a single

157 Ibid., 28–9.

158 Ibid., 31; Kalle Lyytinen, "A Taxonomic Perspective of Information Systems Development: Theoretical Constructions and Recommendations," in *Critical Issues in Information Systems Research*, ed. R. J. Boland Jr. and R. A. Hirschheim (Chichester: John Wiley & Sons, 1987), 3. For information systems research, see, for example, Richard Mason and Ian Mitroff, "A Program for Research on Management Information Systems," *Management Science* 19, (5), (1973): 475–87; R. J. Boland Jr. and R. A. Hirschheim, eds., *Critical Issues in Information Systems Research* (Chichester: John Wiley & Sons, 1987); Wanda Orlikowski and Jack Baroudi, "Studying Information Technology in Organizations: Research Approaches and Assumptions," *Information Systems Research* 2, (1), (1991): 1–28.

159 Macbride, *Automated State*, 79, 82. See also the classic book by Shoshana Zuboff, *In the Age of the Smart Machine: The Future of Work and Power* (Trowbridge: Heinemann, 1988), in which information technology is considered capable not only of imposing information (as in programmed instructions) but also of producing it, thus making its own functioning visible in a new way.

receiver but with many-valued relations among a possibly large number of communicators, not with one-way processes of communication but with interaction and complex circular flows."[160] The approach extended and deepened Shannonian theory beyond its core area, and anchored it to diverse institutional practices that did not seem to have anything to do with the mathematical theory of communication. It was through this tradition, it could be said, that the theoretical legacy of cybernetics and information theory was, to a large extent, solidified as a technology of governance in the framework of management. The organizational structures in question appeared, then, largely as a circular information system, with management information systems as its center. Therefore, the later systems approach as embodied in practices to formalize managerial technologies, systems, and theories was, to a great extent, a continuation of the cybernetic tradition in the social domain, as opposed to some limited expert systems the scope of which Wiener himself hesitated to exceed.

Spreading from military projects into industry and government, with the help of the computer it rationalized and centralized administration in the Department of Defense under Robert McNamara (in the form of the Planning-Programming-Budgeting System), and eventually every federal agency and department, especially through Great Society programs.[161] During the same process, it consolidated authority by eliminating discussion over policies and procurement decisions. This was in line with the way President Herbert Hoover had attempted to plan the economy and coordinate interests by using apolitical, scientific experts as mechanisms to achieve his "associative state," and the way associative institutions were converted into a regulatory state during the New Deal.[162] The Great Society programs utilized the systems approach in seeking to introduce the mission-oriented, not discipline- or department-bound, approach applied with good results in the earlier massive national projects. Yet it was indubitably also the lure of eliminating politics from decision-making that appealed to President Johnson's sense of governance.[163]

It was also during this time that Keynesianism started to gain ground in American politics, although it never came to prevail in the political discourse on the same scale as in Europe.[164] With the increase in influence of the ideas of

160 Klaus Krippendorf, "The Systems Approach to Communication," in *General Systems Theory and Human Communication*, ed. Brent Ruben and John Kim (Rochelle Park: Hayden Book Co., 1975), 158.

161 S. M. Amadae, *Rationalizing Capitalist Democracy: The Cold War Origins of Rational Choice Liberalism* (Chicago: University of Chicago Press, 2003), 57–75; Mintzberg, *Strategic Planning*, 6, 116–17; Hughes, *Rescuing Prometheus*, 141; Waring, *Taylorism Transformed*, 32–5. See also Charles Hitch, "A Planning-Programming-Budgeting System," in *Science, Technology, and Management*, ed. Fremont Kast and James Rosenzweig (New York: McGraw-Hill Book Co., 1963).

162 Hart, *Forged Consensus*, 40–61; Waring, *Taylorism Transformed*, 198; Amadae, *Capitalist Democracy*, 68–71.

163 Hughes, *Rescuing Prometheus*, 167–9. Yet, it is true, the regulatory process also gradually became more open and democratic in the Great Society period (see Horwitz, *Regulatory Reform*, 87–8).

164 e.g., Hart, *Forged Consensus*, 119; Brinkley, *Liberalism*, 47–54, 58–9, 90, 129, 147. One of the key opening chapters in the introduction of Keynesian social planning policy in America was the Full Employment Bill of 1945 (see Elizabeth Fones-Wolf, *Selling Free Enterprise: The Business Assault on Labor and Liberalism, 1945–60* (Urbana: University of Illinois Press, 1994), 21, 33; Brinkley, *Liberalism*, 92). For

John Maynard Keynes, it gradually became evident that the state could manage the economy through fiscal policies without managing the institutions playing a part in it. In spite of the fact that President Roosevelt's advisors were influenced by Keynesian principles, these ideas began to have an effect on American policy-making only after World War II. While widely accepted during the postwar period, Keynesian economics became the premise upon which American fiscal policy was based, and the institutionalization of Keynes came to represent an indispensable element of the Great Society program.[165] After Keynes this approach was combined with neoclassical economics to produce what is generally termed "neoclassical synthesis," which came to dominate mainstream macroeconomic thought. What is of interest at this point is that in the heyday of American Keynesianism the techniques it provided were often viewed, at least among its advocates, as purely neutral: "Keynesian economics is not conservative, liberal, or radical. The techniques of economic stimulation and stabilization are simply neutral administrative tools."[166] It was the same ambition to distance the actual procedures from the political choices that united the common outlook from economic policy to business strategy.

However, what is even more worthy of note is the fact that the techniques and strategies associated with Keynesianism developed into methods of governance that came to be widely used beyond the narrowly understood field of economic policy. According to Mitchell Dean, for example, these techniques, in the form of the welfare state and mechanisms of social insurance, constituted a form of governance in which a feedback loop was established between social and economic processes. In this way, "meeting the needs and managing the risks of citizens" was not a separate problem from, but rather had an effective relation to, "levels of aggregate demand affecting the inverse variations of unemployment and inflation."[167] Another way of putting this is to suggest that governance no longer recognized any pre-given intrinsic boundary lines but was extended to cover the problems concerning citizenship and democracy, *and* those of unemployment, poverty, and inequality as essentially interconnected. Because they fed into one another, they had to be taken care of as part of the same process. This changed view was linked to the new ways of making social reality amenable to government through the newfound scientific approach to government, corporate management, and the conceptualization of society as a cybernetic entity.[168]

As far as business organization was concerned, new fields of knowledge and practice were brought into being. Management science, for instance, was a new discipline emerging largely from the theoretical foundation laid down by

the philosophical aspects of Keynes's economics, see Jochen Runde and Sohei Mizuhara, eds., *The Philosophy of Keynes's Economics: Probability, Uncertainty and Convention* (London: Routledge, 2003).

165 Lekachman, *Keynes*, 270–1.

166 Ibid., 285.

167 Dean Mitchell, *Governmentality: Power and Rule in Modern Society* (London: Sage, 1999), 204.

168 See Mirowski, *Machine Dreams*, 182. Mirowski (*Machine Dreams*, 519) also strongly argues that Keynesianism was closely allied with both econometrics and the spread of the computer.

cybernetics and information theory, on the one hand, and by experiences obtained from the large national projects of the 1950s such as SAGE and Atlas, the first successful intercontinental ballistic missile project, on the other. In time, management science established itself as an independent branch of science distinct from the mechanical models of operations researchers.[169] Yet, the conceptual terrain out of which management science emerged, along with parallel and overlapping discourses on questions of governance, was influenced by the same ethos of control that had permeated a whole range of distinct disciplines. These discourses were closely tied from the start not only to cybernetics but also to the biomechanical metaphor. Let me cite three prominent theorists of the field — namely, Jay W. Forrester, Stafford Beer, and Herbert Simon — each of whom wrote seminal contributions on thought about communication from the angle of computerized control, and at the same time reorganized the conceptual resources inherent in the underlying organicist perspective.

1) For an information system to be practicable, relationships between its parts have to be optimized to minimize the sources of error in the system. This was achieved more and more through a model based on the feedback principle. According to Jay W. Forrester, a pioneer in computer engineering and systems sciences, the feedback principle in information systems is involved when "the environment leads to a decision that results in action which affects the environment and thereby influences future decisions."[170] In other words, feedback concerning the functioning of a system determines the direction of further measures. He launched a program based on a feedback-system perspective for the design of industrial and social systems, which he called "industrial dynamics." Forrester likened the planning of a technical system for desired performance to the planning of policies for social systems, and hence the task of a manager or politician to that of an engineer.[171] Certainly, this program needed the whole history of control of and by means of communication as a precondition. Forrester seemed to combine reliance upon formal methods with the idea of the feedback principle into a consistent system dynamics on the basis of experiences learned from nationwide system projects. In the hands of Forrester, a social system resembled SAGE, with feedback loops being generalized as the hallmark of all social systems. Whereas in SAGE — the predecessor of which Forrester headed (the Whirlwind computer project) — surveillance information was fed into the computer center and compared with preprogrammed models that determined the corrections to be made when needed, in system dynamics a feedback mechanism revealed whether a social project was moving toward the intended goal or whether it needed any adjustments.[172] The

169 Kenneth Boulding, "General Systems Theory: The Skeleton of Science" in *General Systems Theory and Human Communication*, ed. Brent Ruben and John Kim (Rochelle Park: Hayden Book Co., 1975 [1956]), 32.
170 Jay Forrester, *Industrial Dynamics* (Cambridge: MIT Press, 1968 [1961]), 14.
171 Hughes, *Rescuing Prometheus*, 177.
172 Ibid.

essential structure of social systems, being abstracted from complex processes, could thus be modeled, processed, and manipulated by the computer.

2) In the late 1950s it became clear that what was considered as the guiding center of a communication system was, in fact, a whole hierarchical universe of its own. It was seen as being composed of a number of interlinked subsystems, between which information flowed along both vertical and horizontal lines and was processed on several distinct levels of hierarchy. Stafford Beer, one of the pioneers of the information systems tradition, developed a theory of any viable system, based on biological connections. His model consisted of a complex structure of interrelating groupings of several distinct subsystems. Beer systematically elaborated the view of a firm as an organic information system by "tracing the physiology of a control system in terms of a model of the human nervous system."[173] Instead of remaining at the level of neurophysiology, however, he aimed ultimately at moving on to neurocytology in terms of the level of detail to be taken into account.[174] He saw firms, understood as information systems, as instances that collected, stored, processed, and generated quantitative information about the operational world. Starting from the general cybernetic framework of control, he assumed that control is what facilitates the existence and the operation of systems. Yet, unlike in the simple mechanical model of the nervous system, Beer developed the idea that the control function was spread throughout the system, not located in one place. Moreover, and related to the former point, he held that hierarchical control was only one dimension of control, not its only type.[175] These should not be seen as deviations from the classic nervous system model, however, but rather as its more rigorous and detailed application to management theory than before. In fact, it was at Beer's hand that the potential of the biological model was nearly exhausted through a meticulous, scrupulous, even pedantic systematization in the service of management science and the study of social structures in general.

3) This is not to say, however, that the conceptual resources used throughout modern history of communication would be gradually given up. On the contrary, they remained as a long-lasting source for thought, as evidenced by Herbert Simon, one of the most influential social scientists of his generation. Building on Chester Barnard, Simon, who has been called the leading post-Taylorite theorist,[176] developed theories of human and bureaucratic rationality that sought to go beyond scientific management. However, by sticking to the premises of centralized power and specialized operations, he undoubtedly remained within classic Taylorism.[177]

173 Stafford Beer, *Brain of the Firm: The Managerial Cybernetics of Organization* (London: Allen Lane, Penguin Press, 1972), 154.

174 Beer, *Brain*, 258.

175 Ibid., 35, 37, 138. See also Stafford Beer, *Cybernetics and Management* (London: English Universities Press, 1959). For comparatively recent analyses of Beer's ideas, see Andrew Pickering, "Cybernetics and the Mangle: Ashby, Beer and Pask," *Social Studies of Science* 32, (3), (2002): 413–37; Andrew Pickering, "The Science of the Unknowable: Stafford Beer's Cybernetic Informatics," *Kybernetes* 33, (3/4), (2004): 499–521.

176 Waring, *Taylorism Transformed*, 49.

177 Ibid., 77.

According to Simon, the effectiveness of bureaucratic rationality depended on communication of authority. Once successfully set in place, "the organizational values and situation," not "personal motives," determined "the one and only one 'best' decision."[178] Like Barnard, Simon saw organizational communication largely through the idea of the nervous system. Organizations were portrayed as kinds of institutionalized nervous systems that divide, routinize, and bind decision-making process to make it manageable.[179] It was a two-way process in which information flows both to a decision-making center and from it. Although communication process takes place at many organizatorial levels simultaneously, the totality of this mechanism ultimately arranges events into one whole. Simon admits, it is true, that formal communication channels constitute only one part of the total network, but this part is absolutely fundamental: informal communication is only a supplement to a formal one.[180] It was the formal hierarchy of an organization, not unlike that of a nervous system, that determined the nature of communication.

In the later phase of his career, one influenced by developments in cybernetics, computer science, and artificial intelligence, Simon aimed at applying the functioning of the mechanism based on circular structures to human thinking and problem-solving in general. In this project, he used computer metaphors to explain information processes common to both computers and people. According to Simon, both were controlled by the same laws of rationality and, therefore, both were similarly programmable.[181] What this meant for the theory of management was that management appeared essentially as a feedback mechanism. The manager became the central node through which information circulated and by whom communication flows were structured and controlled; it was the responsibility of the manager to ensure that the action taken met the goals fixed. Managers, in this scheme, "were corporate servomechanisms adjusting behavior by communicating deviations from goals," as Waring put it.[182]

In this way, the systems theory, as well as discourses inherently related to it, operated, as did its predecessors, within the framework of corporate capitalism, which determined the position of communication as an object of analysis. Because of this precondition of relevance, systems theory made organizational goals, practices, and relationships appear to be constants inherent in nature.[183] It did not question these relationships, but worked on the basis of them. Rather, they designated what allowed systems science to become a distinct management theory and technique. In this task, it aimed at formalizing decision-making by

178 Cited in ibid., 57.

179 Gareth Morgan, *Images of Organization* (Beverly Hills: Sage, 1986), 81. See also Linda Dennard, "Neo-Darwinism and Simon's Bureaucratic Antihero," *Administration & Society* 26, (4), (1995): 464–87.

180 Herbert Simon, *Administrative Behavior: A Study of Decision-Making Processes in Administrative Organizations* (4th edn). (New York: Free Press, 1997 [1945]), 208–9, 222.

181 In particular, Simon held that mathematical techniques and electronic data processing had together made it possible to automate programmed decision-making and to bring into the programmed area formerly unprogrammed decision processes (see Simon, *Shape of Automation*, 92).

182 Waring, *Taylorism Transformed*, 70. For a discussion about the relationship between Simon and the "cyborg sciences," see also Mirowski, *Machine Dreams*, 452–72.

183 Waring, *Taylorism Transformed*, 26.

using scientific methods and mathematical language. Of course, there was an apparent connection between the increasing use of mathematical tools and the fact that computers could efficiently handle numerical information. Therefore, there was a movement to give ever more multifarious processes and relations an exact mathematical representation. So, while the mathematical form of information made digital computing machines possible, the increasing power of the latter led more and more problems to be analyzed by mathematical tools. What marked discourses during this era of the computer, at any rate, was faith in the applicability of "automation concepts," together with "electronic systems," in administrative control.[184]

Self-Mechanisms of Communication

Establishment of systems theory as the founding framework for discussion of communication and control, and the computer becoming a precondition of modern social life, together laid the foundation for the discontinuity and displacement of the field of communicativity, reflected in both discourses and techniques. A social outlook based on rational planning engendered, along with governance that was becoming more and more computerized, a novel communicative space in which the character and functions of communication were reproblematized within the context of the changes in social conditions and goals. Systems thinking tended to capture all communicative events into their recursive circular movements and thus to defy the ineluctable yet slow process of the disintegration of order and dispersion of communication. They made, it can be said, the social development appear as inherently structured and, thus, controllable by rational means. In this sense, what ultimately became the source of the understanding of communication coherence as an administrative or managerial concept was this perspective of control. This perspective not only was the common thread within the history of communication followed here, forming a way for the theorists to conceptualize communication in terms of a whole, but also forged a relationship between this whole and the limits of communication. Let me finally try to characterize the emerging shape of communicativity in relation to what we have discussed.

The taking shape of several discursive and institutional practices that exploited the heritage largely based on the work of Wiener and Shannon can be seen as having instituted, in diverse ways, a new conception of communication. At the same time it is clear that these distinct theoretical discourses on communication were articulated with each other in the framework of a society that was becoming increasingly computerized. These discourses constituted not only the conceptual terrain where this computerization could be thematized but also, insofar as they resorted more and more to formal systems of representation, the source of quantitative information that encouraged and intensified the very process of computerization. Large national projects as well as business-specific schemes

184 Howard Levin, *Office Work and Automation* (New York: John Wiley & Sons, 1956), 3; George Kozmetsky and Paul Kircher, *Electronic Computers and Management Control* (New York: McGraw-Hill Book Co., 1956), 2.

engendered, besides the technical environment, new discursive forms that buttressed and legitimized similar projects in the future.

If cybernetics and information theory laid down the theoretical framework within which communication was conceptualized, three separate yet related lines of influence, taking shape within this framework, were central to the formation of the new communicative form. While irreducible to any single idea, practice, or technique as such, the convergence and co-development of, first, the emergent science and technology of digital computing; second, the theory and research of communication; and third, the discipline of operations research along with management science and management information systems, set the stage for the new concept of communication and for the corresponding communicative assemblage. Fundamental to this new understanding was the idea of communication as a complex information system, consisting of a number of hierarchical levels and controls within which the functions, meanings, and effects of an individual channel could be identified and analyzed. It was this understanding that came to dominate the communication research.

The origin of modern American communication theory was within the same discursive field in which the cybernetic perspective was introduced to the social sciences. A most important conduit through which this took place was a series of conferences arranged between 1946 and 1953, sponsored by the Josiah Macy Jr. Foundation.[185] Working within this discursive environment, a sociologist named Paul F. Lazarsfeld, who was one of the initial members of the Macy group but who subsequently dropped out, articulated, with Elihu Katz, a sociologist and communication theorist, an approach to communication that laid the theoretical basis for what became American communication research, later renamed as mass communication research.[186] It was the empirical-quantitative approach of this program that formulated the conception of communication that became the standard view at least until the 1970s. The vision and perspective set up here presented mass communication as a new unifying force that reached every single individual.[187] As the leading paradigm under which communication processes were studied was taken from the practices of natural sciences, narrowly defined, it was thought possible to independently measure communicative processes and rather reliably and accurately demonstrate their causal relationships, much as in a laboratory environment. The process of communication was held as being divisible into its smallest natural components and the elements isolated in this way could be named, their respective roles and functions determined, their interrelationships deduced, and the whole process given a cogent scientific basis.

Upon further scrutiny, the treatment communication has been subjected to here does not constitute a break in the way the subject has traditionally been

185 The Macy conferences included mathematicians and physiologists such as Wiener, von Neumann, Warren McCulloch, Walter Pitts, Rafael Lorente de Nó, Arturo Rosenblueth, Julian Bigelow, Gregory Bateson, Margaret Mead, G. E. Hutchinson, Lawrence Kubie, Lawrence K. Frank, Heinrich Klüver, Kurt Lewin, and Paul Lazarsfeld.

186 Heims, "Machine-Organism Analogies," 369. See also Conway and Siegelman, *Dark Hero*, 159, 161.

187 Elihu Katz and Paul F. Lazarsfeld, *Personal Influence* (New York: Free Press, 1955), 16.

approached in scholarly practices. On the contrary, the same type of methodological approach that found sophisticated formulation in Taylor has only been redeveloped here according to modern scientific standards under the influence of cybernetic discourse. If Taylor discovered the irreducible prime particles of a work performance, the standard view of Lazarsfeld and Katz sought a generalized structure for any effective communication process. So, whereas Taylor divided a given work process into its smallest relevant elements, here the same has been done to human communication *in general*, regardless of the context in which it occurs. Thus it also simultaneously extended Wiener's conception of communication while following the same *ethos* and the idea of science.

Yet there is no doubt that it was the elegant Shannon and Weaver model that unified American communication research. To a great extent, the whole field from the 1960s to the 1980s was based on the conception of communication to which Shannon gave an influential theoretical form, and for which he bequeathed the perspective of linear causality. This encouraged subsequent, relatively straightforward research programs, which attempted to identify and isolate the linear impact of one or more variables on another variable. The point to be noticed here is not, however, so much related to the linear logic of the model, or to any single constitutive part of the communicative constellation, but rather to the underlying concept of the explainability of everything relevant to human communication, and particularly media effects that were of interest at the time, becoming common. It opened up a theoretical perspective within which effective relationships could be identified and the structure of a communicative process, once isolated, formalized. Yet, just as theoretical contributions create, enable, and open up, they also dissolve, efface, and conceal. What this orientation missed in particular was the collective nature of communication, the fact that the positions of sender and receiver are not isolated and independent nodes that define the communicative situation, but on the contrary are conceptual entities that are defined already by and in communication.[188] Insofar as communication always originates in something social, the categories of sender and receiver become intelligible only as part of an existing communicative community. It is not accidental either that the theory of functionalism that gained a strong footing in social sciences during that time was criticized for similar reasons.[189]

System approach retained a close relationship to communication systems, which was one cause for *systemicity* emerging as an object of investigation in social sciences. Thus, for instance, Peter Galison, a historian of science, has suggested

188 e.g., Régis Debray, *Media Manifestos: On the Technological Transmission of Cultural Forms*, trans. Eric Rauth (New York: Verso, 1996), 44–5. See also Régis Debray, *Transmitting Culture*, trans. Eric Rauth (New York: Columbia University Press, 2000).

189 Hardt, *Critical Communication*, 14. For functionalism or "structural functionalism," see e.g., Talcott Parsons, *The Structure of Social Action* (New York: Free Press, 1949). For the differences and similarities between Parsonsian sociology and systems theory, see Lilienfeld, *Systems Theory*, 196–9. Looking back to the relationship between control and the systems outlook, Parsons stated later that "[c]larification of the problem of control was immensely promoted by the emergence [. . .] of a new development in general science — namely, cybernetics in its close relation to information theory. It could now be plausibly argued that the basic form of control in action systems was of the cybernetic type" (cited in Heims, *The Cybernetics Group*, 184).

that the impact of the system perspective on social sciences had its origin in the turn-of-the-century networks of telephony and power.[190] While it was in the context of these networks that the idea of systemicity first started to take shape in social thought, the real breakthrough did not take place until in the 1940s and 1950s. The Macy conferences held at that time introduced the cybernetic perspective, with its inherent relationship to communication engineering, to social sciences. If the notion of system did find inroads into the conceptual framing of broader social questions independently of systems sciences, it was precisely the perspective adopted by cybernetics, however, that subsequently came to permeate a wide range of social sciences and became used in contemporary social and political rhetoric. During this process, a system lost its static character of hierarchically arranged elements with an intrinsic nature: it became a dynamic concept.

The vision of social processes as a big information structure was not limited to communication research. While this research largely adopted the linear interpretation of the Shannon and Weaver model, a number of different disciplines from organizational theory to political theory and social psychology started to analyze social groups as communications systems characterized by a multiplicity of feedback loops.[191] Moreover, along with the shift within the cybernetic movement itself, many theorists influenced by the ideas being developed in this movement turned to the concept of feedback instead of to that of homeostasis, as already mentioned. What the diverse theorists and disciplines shared was the same information-processing metaphor that differentiated a system from its outside by means of an inbuilt mechanism of control.[192]

In the process of cybernation of communication, so to speak, new metaphors replaced the mechanical metaphors through which it used to be conceived. Though they bore traces of mechanical technology, too, what was essential was that they exploited the new world of computing technology in that the relationships they conceptualized were logical, not physical. The world portrayed by the systems approach was based on a view of the mutual internal relationships of nodes of communication, the overall structure of the continuous exchange of these dispatched messages being organized by the principle of self-regulation. This view of a system correcting its own functioning took the place of the previously held image of rigid mechanical performance, blind to surrounding environmental changes. Gradually leaving mechanical metaphors, communication now entered the sphere of influence of information system analogies.

The inevitable consequence of recourse to formal systems is that the more the technical aspect was emphasized, the more urgent it became to integrate the

190 Galison, "Ontology," 256. Galison, however, omits the telegraph from his remark. Menahem Blondheim corrects this by arguing that the system approach, implying the fact that tools and machines could only be used in integrated systems, was first made manifest in the telegraph and the railway (Menahem Blondheim, *News over the Wires: The Telegraph and the Flow of Public Information in America, 1844–1897* (Cambridge, MA: Harvard University Press, 1994), 55–6).

191 Dechert, "Development of Cybernetics," 20.

192 e.g., Deutsch, "Mechanism, Organism"; Deutsch, "Mechanism, Teleology."

human element into the system as component part as it implied a factor of insecurity. To control this element, the machine metaphor still provided a useful tool. On the other hand, it was within this very conceptual framework that human behavior became seen as problematic in the first place. In order to overcome this strange situation, there was an inclination — especially among systems scientists and practitioners — to emphasize an unambiguous formalism, coupled with mathematical tools.

The reformulated machine metaphor — that is, the information system metaphor — provided a model for thinking about and planning the social processes in a time when the influence of computers began to reverberate in all social activities. Especially, the idea of a computer system played an important role in both understanding and organizing the social realm as the object of governance. In fact, the computer system, it was thought, tended to determine not only the functioning of socioeconomic planning systems but also the form.[193] This is not, of course, surprising, as many of the key concepts that have come to account and symbolize some central aspect of society can be traced back to technological systems, as Galison seemed to think. Like the telephone system earlier, the computer system seemed to provide a model for the entire structure of society.[194] The computer, then, was used to build, in the words of an air force colonel, "a servomechanism spread out over an area comparable to the whole American continent."[195]

As for organizations, the information systems view had a powerful effect, as many organizations increasingly came to rely on information systems.[196] In the 1960s, Ulric Neisser, an American psychologist, grasped the profound influence that the metaphoric use of computer vocabulary had come to have in public intercourse. By referring to the impact of cybernetics in several analogies such as "the brain is like a computer," "man is like a machine," and "society is like a feedback system," he brought out the increasing role of the computer in contemporary sentiment.[197] It is clear that the displacement in the communicative space did not imply just some adjustments in technological development or in theoretical thinking: it changed the scope of what was possible and what was not and in so doing redefined the ontological presuppositions of the age. Thus, the computer and related discourses and practices did not operate only on an instrumental level. Rather, they introduced novel resources for collective self-understanding. The perspective of information systems seen as feedback structures was not only fairly coherent but also very widely adopted. It constituted the conceptual framework for many contemporaries through which they could explain to themselves what

193 Macbride, *Automated State*, 154.

194 This is not to say, of course, that the telephone or the computer system would have been, respectively, the sole or privileged sources of social self-understanding and self-representation. Naturally there were a number of other sources that served this purpose. Yet models related to these systems were no doubt absolutely crucial for contemporary discourses concerning the state of society.

195 Cited in Joseph Weizenbaum, "Computers, Tools, and Human Reason," in *Communication in History: Technology, Culture, Society*, ed. David Crowley and Paul Heyer (New York: Longman, 1991), 280.

196 Morgan, *Images*, 83–4.

197 Ulric Neisser, "Computers as Tools and as Metaphors," in *The Social Impact of Cybernetics*, ed. Charles Dechert (Notre Dame: University of Notre Dame Press, 1966), 72.

was going on in their societies. Jay Forrester encapsulated this view by proclaiming that "everything we do as an individual, as an industry, or as a society is done in the context of an information-feedback system."[198] The vision of an information system had thus developed into a form of intelligibility that established the setting for understanding the meaningfulness of key social processes.

If at the beginning of the cybernetic line of research, information handling became a more central problem to systems analysis than handling of energy, now questions relating to transmitting, storing, and processing information began to be folded ever more directly around the social organization. They were seen as properties characterizing profoundly *the social* as such. In this framework every system was treated as an information system that opened up a whole new territory for methods of control. The convergence of computer technology and systems-scientific methodologies seemed to finally provide the necessary means for overcoming politics in governing the social system. "Given the resources of modern technology and planning techniques," stated Macbride, "it is really no great trick to transform even a country like ours into a smoothly running corporation where every detail of life is a mechanical function to be taken care of."[199]

This perspective provided, then, the new whole within which communication became conceptualized. It had four main characteristics. It had the capacity to, first, sense and monitor its environment; second, relate this information to the operating norms that guide system behavior; third, detect deviations from these norms; and, finally, initiate corrective action when discrepancies were detected.[200] What is of interest here is that communication became a function of this whole, which defined its nature and its identity. It could not have been thought of independently of this whole, since it was only because of its functionality that communication was given its distinctiveness as an object and means of control.

Thinking about communication has always been dependent on a totality that has been the source of its identity. While the theme of control began to dominate the context through which communication came to be understood, the idea of systemicity formed the channel that connected it to the sphere of control. This was no longer the mechanical systemicity of the past, but something that could be managed by numbers, as Waring put it.[201] It was exactly the logical nature of communication revealed by and within mathematical formulas that established it as a control function. Mathematical relationships, as the rational basis for control, thus became the immanent relationships of communication itself.

This is also a conjunction that helps us to see the relationship of the systems approach to the question of the limits of communication, which continued to be conceived in the framework of an original whole. Systems thinking conceived communication as a function that was constitutive of the whole it belonged to

198 Forrester, *Industrial Dynamics*, 15. For an insightful and careful account of the language of information systems and information processing in social thought, and especially in economics, see Mirowski, *Machine Dreams*.

199 Macbride, *Automated State*, 187.

200 Morgan, *Images*, 86–7.

201 Waring, *Taylorism Transformed*, 20.

(nervous system, machine, community). Within this tradition, the whole of communication was always an immanent totality because it was self-sufficient and did not presuppose other relationships in order to be understood. This mechanism, of course, implied the existence of something other — the "environment" — but this other assumed its significance from the immanence of communication itself. For this reason, all the new departures relating to bringing communicative relationships themselves to the fore, instead of focusing on phenomena separated from one another, must be seen as subordinate to the concept of dominant totality. In fact, although it is correct to ascribe thought about communication itself to the systems approach, it was only after the dissolution of the formalistic program that communication could be thought of as independent of any constituting whole.[202] While the thinking on communication has, in a certain crucial sense, broken away from preset frames, the *network* seems to have become the new generic form of the communicative whole. This development was linked with many contemporaneous processes, including the spread of microcomputers and computer networks — perhaps the most famous being created by the Defense Department's Advanced Research Project Agency — and the proliferation of instances and relationships of communication, leading to a change in the whole communicative formation.[203]

Generally speaking, an increasing tendency toward decentralized systems seemed to point the way for the future of communication. System-theoretical ideals lost the position they held when participation in decision-making started to assume an increasingly important role in organizations, with the subsequent demands to open up communication channels and to diffuse authority.[204] Large system projects since the 1960s, along with the development of computer networking and the launching of ARPANET, changed decisively the way complicated processes and dynamic organizations were approached. In 1969, the Planning-Programming-Budgeting System was officially discontinued in federal bureaucracies due to problems related to its hierarchical and centralized decision techniques.[205] Moreover, in the 1970s the technocratic strand of management science lost much of its former appeal because of the American manufacturing crisis, and came largely to be replaced by "soft" techniques developed in the form of soft operations research by Russell L. Ackoff and C. West Churchman in the United States and appreciative systems theory by Charles Geoffrey Vickers in Britain. This turn implied a leaning toward working with people's cognitive processes, which were customarily looked upon with suspicion in "hard" systems thinking and were therefore excluded.[206]

202 The dissolution of the formalistic program is here associated with the end of the classical cybernetic approach of the late 1960s and early 1970s.

203 e.g., John Thompson, *The Media and Modernity: A Social Theory of the Media* (Cambridge: Polity Press, 1995); Neuman, McKnight, and Solomon, *Gordian Knot*; Andrew Shapiro, *The Control Revolution* (New York: PublicAffairs, 1999); Asa Briggs and Peter Burke, *A Social History of the Media: From Gutenberg to the Internet* (Cambridge: Polity Press, 2002).

204 Wren, *Management Thought*, 334.

205 Amadae, *Capitalist Democracy*, 71.

206 Waring, *Taylorism Transformed*, 35–42; Peter Checkland, "Systems Theory and Management

According to this new interpretation, hard mechanisms aimed at producing efficiency only veiled disagreements about values and goals that were never eliminated and, therefore, the inevitable political nature of governance. Thomas Hughes has depicted the new, networked communicative form that emerged, from an administrative point of view, as including

> a flat, collegial, meritocratic management style as contrasted with a vertical, hierarchical one; the resort to transdisciplinary teams of engineers, scientists, and managers in contrast to reliance on discipline-bound experts; the combining of diverse, or heterogeneous, physical components in a networked system instead of standardized, interchangeable ones in an assembly line; and a commitment by industry to change-generating projects rather than to long-lived processes.[207]

Basically, what is at play here is a change in the way a communication system worked. Whereas these systems, from the telegraph to telephony, were traditionally built top down, new communication systems, by contrast, were constructed bottom up.[208]

Hand in hand with the development of new information and communication technology, with its undermining effects on hierarchical structures now viewed as rigid and inflexible, a new antihierarchical sentiment emerged. This cannot, of course, be called a direct causal relationship, for whereas the new technical systems and organizational forms based on them made up an object and a place for discursive investigations, these discourses generated new institutional practices and structures too. It was, nevertheless, at this stage that the fundamental transformation in the images and metaphors from hierarchical to more horizontal communication occurred. This was largely because of two converging trends. First, rigid corporate hierarchies were replaced by more flexible systems emphasizing horizontal relationships. Second, improvements in information and communications technology seemed to facilitate decentralization while maintaining managerial control.[209] As mentioned earlier, the horizontal aspect has never been missing from the technical history of communication, but during this period it certainly come to the fore and paved the way for the concept of networked communicativity. On the other hand, centralized power hierarchies, although entering into a new era, did not vanish either. It is rather a question of a new power distribution in a constellation that facilitated horizontal relationships more and with different intensity than previously was the case. At the same time, it opened up the possibility to establish the kind of network-based relationship to the world

Thinking," *American Behavioral Scientist* 38, (1), (1994): 79–90; Peter Checkland, *Systems Thinking, Systems Practice* (Chichester: John Wiley & Sons, 1999), A49; Robert Flood, *Rethinking The Fifth Discipline: Learning within the Unknowable* (London: Routledge, 1999), 51.

207 Hughes, *Rescuing Prometheus*, 5.

208 Lebow, *Information Highways*, 189. See also G. J. Mulgan, *Communication and Control: Networks and the New Economies of Communication* (Cambridge: Polity Press, 1991).

209 Waring, *Taylorism Transformed*, 188. See also James Cortada, *Strategic Data Processing: Considerations for Management* (Englewood Cliffs: Prentice-Hall, 1984); Thomas Valovic, *Corporate Networks: The Strategic Use of Telecommunications* (Boston: Artech House, 1993).

that broke through the preceding images and resources for self-understanding they provided.

Insofar as occidental history has been closely tied to the formulation of ever more precise definitions and formal analytical frameworks, it is inevitable also about the multiplication of these forms and practices and the emergence of novel, mutually incompatible, and alternative discourses. This is exactly what Gianni Vattimo referred to when speaking about the birth of new centers of history in his *La società transparente*.[210] The contemporary communicative environment consists of distinct overlapping and parallel strata that cannot be unified under any single conceptual principle: current networks are in a state of constant alteration with continuously shifting boundaries. Although discourses have perhaps become more technical, the processes of technicization are numerous and heterogeneous.

The line of development in communications that could be construed as a comprehensible sequence from the early telegraph to the rise of microelectronics has weakened in the age of information networks. This happened apparently because the notion of a network that has come to dominate the representations of society is not a resource similar to the metaphors, conceptions, and images that previously characterized the modern history of communication. On the contrary, it is more porous; it is essentially open in a way that seeks to prevent closure — a closed interior.[211] This is why, in the age of networks, technology and the attendant images and metaphors forfeit their former consistency and can no longer provide any single, totalizing model for society. Although centrally directed systems did not lose their social position — rather the contrary — they could no longer furnish a conceptual model for thinking and speaking about society. This state of affairs did not, however, prevent continuing use of the biomechanical vocabulary. Thus, once switches are computerized and transmission digitalized, all elements of a communication system belong to the same digital circuit, forming nothing other than a "digital nervous system."[212] "Network" is the figure through which we can conceive our society today, and it both resists the presupposition of a closed system and provides elements to reassemble the society under one commonly recognizable image.

However, if it became unthinkable to depict society by means of an exclusive communicative model, the identity of communication itself underwent a radical transformation and lost its consistent nature. During the eventual dissolution of the systems-scientific project, at the same time that information technology and microelectronics became common, the lucid and positive nature of communication gave way to its infinite transformability and flexibility. With this development, the hopes for a unified science of communication that arose in the 1970s from general systems theory proved unrealistic. Seen from this perspective, communication could not retain any specific organizational characteristics, such as those referred

210 Gianni Vattimo, *The Transparent Society*, trans. David Webb (Polity Press, 1992).

211 Kai Eriksson, "On the Ontology of Networks," *Communication and Critical/Cultural Studies* 2, (4), (2005): 305–23; Kai Eriksson, "Foucault, Deleuze, and the Ontology of Networks," *European Legacy* 10, (6), (2005): 595–610.

212 Edwards, *Closed World*, 195.

to in biomechanical metaphors, and instead assumed innumerable convertible and variable qualities, not reducible to any single structure or function. Endless reconstructability and diversity, together with the emerging network-based outlook, were themes pointing the direction for the development of communication. Communication has become our general environment, in which we can articulate, after having become, prior to it, already articulated ourselves. In this way a new opportunity is presented for the thinking of community. It is not necessary any longer to be considered in terms of a never attainable unity, but rather as a network that simultaneously unifies as well as separates, not as a being or goal, but as the structureless structure of being.

5

Communication and Modernity

Although inevitably schematic, this study has tried to bring out the essential points of intersection in the history of the relationship between communication and telecommunication. It is through these nodes and constellations that distinct chains of command, scientific discourses, managerial theories and practices, as well as telephone lines and digital circuits can all be conceived of as an integral part of the history of governance of and by "communication." What unified this path of development was the idea that questions of communication could be seen as problems to be addressed in terms of political and technological strategies. While manifesting itself as an object and means of governance, communication created new forms and images of society which, in turn, enabled new methods of governance.

It is possible to conceive, as we have done here in this study, of the various vocabularies operating in different registers of speech, from the emergent analogy of organism to sophisticated mathematical models, as historically locatable attempts to define "communication," its essence, and limits. Surely, the clumsy biological metaphor, while in its infancy, articulated a different "communication" and its social occurrence than did the later formal theories and models of research, emerging from the institutionalized study of communication in its various forms. All the same, they had without doubt the same object and aim: to seek to understand, by means of the conceptual tools at hand and within prevailing regimes of cultural practices, the core of what "communication" was about. In other words, they sought to capture the essential characteristic of what had come to be thought of as communication, in whose formation the new communication systems played a crucial part. It is also clear that many of these conceptualizations of communication were not independent of one another but, rather, mutually articulated.

It is the metaphor of the nervous system that has served as the central conceptual framework for understanding and representing communication throughout the age of telecommunications. It gave a name to the whole that brought a certain unity to the different mechanisms and constellations of communication. The nervous system named and specified the figure that made the dissimilar practices, projects, and ideas in the final analysis intelligible with regard to the image of modern society: it tied the communicative constellation into a social whole. Although the nature of this whole did not remain the same but changed according to changes and displacements in and of communication from an organic to a feedback structure, there is a certain consistency underlying the vicissitudes of communication.

The notion of a nervous system retained the idea of unification, widely articulated during the introduction of the first nationwide telegraph systems. This idea did not disappear; rather, the concept of a nervous system connected the idea of unification to a powerful vision that made it possible to analyze it theoretically in a more detailed manner than was previously possible. It also allowed this idea to be put in a context in which the technical, political, and economic conditions of unification could be viewed and studied as one whole. Therefore the nervous system must be seen rather as a new conceptual form through which the idea of unification found its expression: it served as both the juncture in which the need for unity was met with a theoretically powerful way of conceptualization and the point of identification of a discursive practice through which this need was continuously rearticulated. It interpreted the whole of communication in a way that met the needs of both conceptual consistency and political governance.

However, while the nervous system has constituted the whole within which communication could be formalized, it has also become formalized itself in the course of this process, and gained a more and more precisely articulated structure in a scientific sense. In other words, the nervous system, being increasingly formalized, has formed the whole within which communication has become both the object and means of administrative action, with the administrative point of view being internal to the modern history of communication, although surely this history can be written from another perspective as well. In it, communication was positioned as a focus for both systematic investigation and political projects, building up an interconnected body of theories, technologies and institutional settings. Yet at the same time, during this development within which communication has been formalized, ever more accurate mathematical methods have come to represent the nervous system. All in all, thought about communication is, in the final analysis, dependent on the formalization of the whole that is its precondition. Thus concurrently with the course of the formalization of communication, thinking about the "entity" of communication has been formalized too, as it is in the end — in this tradition — nothing external to the formalization of the whole designated by the nervous system.

Communication, within this history, has always been closely tied to an anterior whole, from which it has not been able to dissociate itself. This whole was characterized in the age of the telegraph by its naturality, in the era of the telephone by its mechanicality and when entering the age of computer networks, its digital nature (communication and community understood as an "organism," "machine" and "cybernetic system" respectively). Far from constituting coherent systems, these interconnected historical formations however gained consistency through new technologies, shared conceptions and popular images. Once formalized and codified in formal procedures, routines, and designs, this origin, then, has drawn the line that distinguishes communicative relations from noncommunicative ones. However, this cannot be conceived of as a restriction or as some kind of shortcoming because the communicative whole with its demarcations between inside and outside is what has made it possible to pose communication as a theoretical problem, as well as being an outcome of this very thinking: without boundaries there would be no meanings. At the same time, it has enabled systematic regulation of the interrelationships between the system and its outside in many different situations, influencing the practical design of communication systems in many local contexts, which have been the points of origin for certain general communicative techniques, discourses, and practices.

The tradition studied here illustrates that the process in which communication has become a distinct political and theoretical concern is inherently entwined with the history of the "ontologization" of communication. Communication can be viewed as taking its place as a relationship between the occurrence and withdrawal of instances of it, on the one hand, and its ontological conditions, on the other. Communicative manifestations and relationships bring something into being; they appear in communication but are not communication itself. On the one hand, this communication in "itself" cannot be expressed in communication, because a manifestation of communication is not capable of expressing the fact that there *is* communication, but, on the other hand, in order to "work" communication presupposes some conception of the whole of communication — in other words, the representation and objectification of communication. This objectification, essential for the functioning of a communication system, is manifested in the history of telecommunication in two related registers that constitute each other. While communication needs conceptual, necessarily reductive descriptions that represent it as a "whole" ("information system"), it also requires technical systems with the attendant quantification of communication for its practical fulfillment ("transmission capacity"). Both of these ways of objectification are inherently linked together so that even though it is possible to analytically distinguish between "conceptual" and "technical" elements, in a given historical process they are indistinguishably tied together. The nervous system has provided a concept that, throughout the history of communication followed in this study, has amalgamated these aspects into a concrete whole. This continuous oscillation between the necessity to objectify communication, on the one hand, and the fact that while being objectified communication is not expressed in "itself," on the other, has determined the scope of communication.

Both conceptual and material entities from organic to cybernetic wholes have enabled communication to be thought of it itself but has also set limits for its understanding. Communication, in the tradition studied here, was a function of a whole constructed as a hierarchy of channels with a control unit organizing the flow of incoming and outgoing messages. Moreover, it was mainly a formal function in that the type of communication regarded as constitutive of the whole was determined through formal lines of communication. Finally, what was fundamental was its repetitive character, not any single meaning-content as such. Although constitutive of the whole, communication, at the same time, depended on it, because it was only within this pre-given totality that communication could emerge as communication. It integrated the plenitude of diverse events and relationships into a conceptual field and allowed them to be seen as controlled by some underlying laws amenable to formalization. This is why the history of modern communication has always had a close relation to the politics of integration and unification.

During the periods marked by the telegraph and the telephone, respectively, the conception of communication was conveyed through a strong instrumental sentiment: whether communication was understood as neutral or as positive, it was in any case a vehicle for prosecuting communal interests. Because these interests were thought to be presentable on behalf of *the common* — which amounts to saying that a social consensus, based on an objective common good, was perceived as possible — communication lent itself naturally to the service of the realization of this objective. Thus, it was possible to continue to consider the community in terms of a unity (if not already of an accomplished unity, then at least of a reachable one): it could be presented by the word "we," which at the same time refers to the subject of the social history.

While embodying the alliance between community and commonness, the rationalization of communication was not an isolated development in modernity. Instead, it represented and reflected a more general principle, the ontological condition of our modern existence, which Heidegger termed *Gestell*, translated into French by Michel Haar as *dispositif* — a term used also by Foucault, albeit in a somewhat different sense. According to Haar, "science that pursues the mathematization of nature is not an autonomous project. It reaches conclusions beforehand about what the real is; it only admits of the objectifiable and calculable. It is in the service of the more general project of technological enframing (Gestell)."[1] Thus, the drive to render communication quantitative, seen from this perspective, can be said of having come about because of an already established world-relation. What made this project so consistent is that the key figures and discourses belonging to it operated firmly within the limits of this relationship, simultaneously articulating its nature and scope in terms of communication. The history followed here, as well as scientific research and technological development as inherent parts of the history of communication, is therefore indistinguishable from this project that characterizes the whole of modernity.

The history of administration of and by communication has, since the beginning of the twentieth century, been closely tied to the idea of total control. This is not to say that the history studied here is alone in its connection to this idea. Naturally, many other systems, typically of a technological nature, have been connected to the idea of total administration, so that this idea has been expressed in and through these systems and related practices. Yet communication systems and related discourses constitute a central part of this history. Now, it is possible to conceive of Heidegger as a thinker relevant to the modern processes of communication, as his thinking was themed, especially during his later period, largely in relation to the idea of generalized technical control. Insofar as he can be considered as having thought about communication's modern lines of governance, these

1 Michel Haar, *The Song of the Earth: Heidegger and the Grounds of the History of Being*, trans. Reginald Lilly (Bloomington: Indiana University Press, 1993), 81.

lines at the same time set the limits within which this thinking could operate as a contemporary critique or as a philosophical analysis of modernity. Heidegger was preoccupied with, even bound to the idea of, centralized power, and the critique of technological dominance is dependent on the presupposition of the idea of a oneness of that power.

Seen in this context, communication has been inherently connected to the history of control. This is so not least because of the fact that it was largely in the sphere of institutional intercommunication that the systems, discourses, and images of communication that were subsequently generalized and established culturally were brought into being. The threats and hopes invested in these were the main object of commentary by contemporary intellectuals. This was emphasized with the improvement of communications when the technological "system" seemed to assume a strongly consistent and even autonomous character. Although clearly Heidegger's thinking cannot be reduced to it, the very idea of this total and unitary system constituted the intellectual context of thinking on technology for Heidegger and for his contemporaries.[2] Subsequent developments and especially the advent of computer networks altered, however, this general pattern and undermined the concept of the dominating center. If ontology has been commonly associated in philosophical thought with consistency and stability, now the era of networks has both scattered and historicized this combination. It seems that thought about community cannot rely any longer on the idea of an unitary system, strong center, method, or law, or any single principle aiming to capture what is going on in communication.

The Fate of Communication

Let me now complete this book by presenting some general insights on communication in the modern period. What characterized the tradition followed here was an emphasis on a common code, form, or method, a contract perhaps, that maintained and protected meaningful communication from misunderstanding or distortion. It was this common order that set the limits of communication by excluding the multiplicity that always risks blanketing it out. It is here that the shared form was protected at the expense of what was purely idiosyncratic by virtue of a mechanism of exclusion: it eliminated what could not be shared, the "irrational" part of communication — in other words, noise. The form of communication, thus, is primary and noise, from the point of view of communication's functioning, a disturbance. At the same time, however, communication cannot do without noise, because it is an inherent quality of the substance in and through which communication takes place; that is, it is the empirical part of communication. Noise, therefore, is essential to the possibility of communication.

As such, noise is the absolute prerequisite to communication and the place in which multiplicity finds its expression. It is not something that would or should be abandoned in order to agree common ground, but rather the origin of all

2 See John McCormick, *Carl Schmitt's Critique of Liberalism: Against Politics as Technology* (Cambridge: Cambridge University Press, 1997), especially chapter 1.

communication. It exposes communication to interference and disorder — to contention, it is true, but without this communication would not be possible. Communication is about constant exposure to what is undividable in noise, which incessantly threatens to swallow it up. This is why any successful communication, of necessity, is about excluding noise. As noise remains an integral part of communication, this exclusion both takes place in and is directed at communication: it is here that communication is born as communication. When succesful, communication finds equilibrium between the noise that all the time threatens to overwhelm it and the commonly agreed form that would enable accord. Communication cannot totally remove noise because it is its precondition, but what it can do is to optimize the relationship between a multiplicity and a unity in a manner conducive to continuous communication. For this it needs a method immanent to the system of communication, the task of which it is to maintain this equilibrium. To establish this kind of mechanism, however, requires the formalization of communication.

Michel Serres has noted that formalization is a process whereby one can move from noise to consensus, from contention to unanimity, or from politics to technology. Formalization, here, is about optimal elimination of noise, not about its complete eradication. Nevertheless, it breaks the relationship with the historical context of its origin, because for the common form the particular context in which it manifests itself is not important. Moreover, it is formalization and the ensuing elimination of the empirical that enables science to be created. Of course, a common form or communication is possible only empirically, and cannot be separated in its appearance from the practices, systems, and techniques in which it manifests itself. Yet communication does not reduce to the empirical, but presupposes something common that exceeds its particular forms of realization; and it is not possible to recognize what is common without excluding the very empirical it relies on.

As we move from the early analogies of a nervous system to a systematic modeling of communication, the link with the historicity of communication is continuously reconstructed. Whereas the first biomechanical analogies sought to conceptualize experience of the prevailing understanding of the social, the description and explanation of communication by means of mathematical tools presenting themselves as objective and apolitical that continued to be carried out within the framework of the nervous system constituted a field for thinking about the historicity of communication in a decidedly more technical way. Of course, this shift is inseparably interlinked with the development and professionalization of the scientific estate, and with its opening up of new research practices and object definitions for the academic institution, spearheaded by modern natural sciences. This approach created the basis for the apprehension of communication in a *methodical* framework and for relating it to the question of efficient organization of action. It is in this sense that thought about the conditions of communication as a technical question corresponds to and expresses the prevailing experience of the social. But at the same time this historical experience conflicts with the fundamental historicity of the social.

On the other hand, these processes of formalization did not emerge outside communication. On the contrary, they came into being within the very communicative networks that they sought to formalize. Communication cannot be separated either from the whole that makes it communication or from the technical system that embodies this whole in a practical organization. Yet it is not reduced to these wholes and systems; they are no more "real" than communication, and, in fact, "are" only because of communication. The scope of communication is thus defined as a continuous and indeterminate relationship between the "absolute structure" of communication, on the one hand, and the empirical forms assumed by it, on the other.

Communication has always operated here as a tool for cutting across, eliminating, and ignoring conflicts and disputes. This is one apparent thematic line that integrates the history traced here. Communication has performed the task of undoing politics, although it is itself nothing other than politics, in that it can be understood only on the basis of politics. If politics constitutes community, then it can be said that communication, in this line of development, was also a mechanism for undoing community as community. That is, it undid community as a social whole characterized by endless negotiations and discussions concerning the coexistence of people. In its place, it laid a regimented whole that was based on constellations consisting of codified relationships, practices, and systems of communication.

I must emphasize, however, that there is no single figure for a given historical communicative constellation. Rather, what is at issue here is an attempt to localize and specify the *telos* of communication at work in each constellation: that is, the aim and goal of its formulation and use, the projects it serves, and the mode of being it enables and supports. It is not a question of totalizing descriptions of particular concepts and practices, but rather of the way by which, to use Mitchell Dean's formulation, "various programmes of conduct and regimes of practices have enfolded within them definite teleologies."[3] Thus, although distinct communicative constellations have "definite teleologies" that allow us to draw a line between them and consider them as having certain individual consistency, and although the nervous system no doubt has operated as a conceptual model enabling this consistency in each case, the communicative "wholes" should be understood not as general terms but rather as names for related regimes of discursive practices and fields of technology. Therefore there is nothing original or essential behind a communicative constellation, no ultimate sense that could explain it all. As Jean-Luc Nancy says, "at the point where we would expect 'something,' [. . .] there is nothing but the manner."[4] What is "behind" this constellation can only be imagined, and the history followed here is exactly the history of this kind of imagining. Far from

3 Mitchell Dean, "Foucault, Government and the Enfolding of Authority," in *Foucault and Political Reason: Liberalism, Neo-Liberalism and Rationalities of Government*, ed. Andrew Barry, Thomas Osborne, and Nikolas Rose (London: UCL Press, 1996), 225.

4 Jean-Luc Nancy, *Being Singular Plural*, trans. Robert Richardson and Anne O'Byrne (Stanford: Stanford University Press, 2000), 14.

being fictitious, however, this imagining is very real and absolutely necessary, *life-preserving*, to use Nietzsche's terminology.

Now, it was not until after the appeal of cybernetics and information theory started to vanish that there occurred a kind of "French Revolution" in thought about communication. Whereas previously communication was conceived of from the perspective of good governance, it now began to be regarded as a unit that governs itself as control became immanent to communication. Yet these two perspectives on communication are not necessarily separated from each other in a clear-cut manner. In fact, the whole history of communication, where it is seen as both an object and a means of governance, can be depicted in terms of steersmanship, or the "science of piloting," from which cybernetics obtained its etymological roots. It is indissolubly tied to concepts of command and control, but these have to be understood in the marine context in which there is no possibility of withdrawing to a private place as one can do on land. Captain, crew, and passengers, taken together, constitute a social whole that is based on constant coexistence. As Serres has pointed out, this community is closed behind a line drawn by the safety fence, and the only option outside the rope is drowning.[5] This is why the governance put into practice is necessarily self-governance, too. It deals with the course and actions to be taken within a dynamic community sharing the same goal. So, work in Bethlehem Steel Company or AT&T was not suspended while it was rationalized according to Taylor's program, and the air defense system was not made idle during its automation. For this reason, too, the telephone system could be seen as the model for self-governance under circumstances of constant change. Here, as in a ship, there is no place to withdraw to from the unceasing responsibility for the functioning of the whole.

The methods used, however, especially if followed to their end as a principle, do not conceal their nature as formal, general rules of steersmanship. This is what makes the modern history of communication, ultimately, the history of control. In a closed community as in a ship, it is of foremost importance to eliminate any misunderstanding or ambiguity over the course of action. What is essential is to do away with any possibility for negotiation whatsoever. This calls for a system of a perfectly understood set of rules and routines of action that stabilize complex activities into a uniform system under the command of a supervisor. In many organizations with clearly defined goals, formal methods for setting up lines of command and the respective relationships among actors were thus absolutely necessary.

It is true that the environments these formal methods were applied to did not resemble a ship in any literal sense but had a much more open and systemic nature. They were, however, the loci in which the means and structure of intercommunication became of central importance. It was in these very settings that communication began to be systematically thought of as a fundamental ingredient of a given action. In addition, from these local and specific contexts emerged the

5 Michel Serres, *Le contrat naturel*, Éditions François Bourin, 1990 (*The Natural Contract*, trans. Elizabeth MacArthur and William Paulson, Ann Arbor, MI: Michigan University Press, 1995).

theories, models, and systems of communication that spread throughout society and were generalized to be applicable to various different yet ever-broader environments outside their sites of origin. Wiener understood this well, which makes him, for my purposes, theoretically the most interesting figure at the end of the line of development followed here.

At the same time, it is clear that because of these reasons communication, insofar as it has been considered within the sphere of systematic thinking, has been thought of on the basis of an origin, of an original whole. Typically, communication systems, along with the various discursive practices and metaphors with which they have always developed in tandem, have served as this type of privileged whole. As a result, this privileged essence has distinguished communication from what does not belong to it, encapsulating the criteria for demarcation. Here, the demarcation stem from communication itself. Consequently, discourses of communication are usually practices of exclusion, in and through which the mode of communicativity proper is constantly engendered and its boundaries monitored and maintained — *within* communication. Whether it is a question of incessant negotiation, signal distortion, or increasing disorganization, the operation of exclusion is always the same, as is the aim of this operation; namely, to strengthen the communicative whole and to give it an identity of its own.

Common to all the thinkers and theorists discussed in previous chapters is the fact that they conceived of communication as a rationalization process or a technique for that end, a technique that was perfected with the help of scientific method. The main stages and experimental scenes of this rationalization were above all modern business organizations and later also governmental institutions. For this reason, it was natural for the thinkers discussed to view communication from the point of view of the objectives and functions of these organizations. The business-driven approach was therefore inbuilt in rationalization of communication becoming an issue in the first place. Yet this is also the reason that connects these thinkers to the tradition of metaphysics, where communication does not appear as a precondition of identities and phenomena but a determinable phenomenon in itself. Because the aim of many of these thinkers was to optimize a system's functionality by formulating its intrinsic scientific laws, they remained stuck within this functionality. This is the very modern fate of communication that cannot be disentangled from the history of communication that has been my concern here. The lineage from Sumner to Simon and from Ward to Wiener can be seen as representing and epitomizing what have been the most influential and lasting elements in the formation of thinking about communication in the modern period, but the problems concerning functionality applies to contemporary thinkers as well.

The attempt to exclude the actual multiplicity and variety from communication must be seen also as an attempt to think about the *limits* of communication. This is not only because the formalization of communication is an integral part of the history of communication. In other words, attempts to represent its intrinsic structure were necessarily constructions in which the limits of communication, variously understood, became constantly rearticulated. However, it is also because the formalized structure of communication was conceived as a *universal* form that would underlie all meaningful communication processes irrespective of their variable contents and historical contexts.[6] It captured in a coherent form what was the essential structure of all thinkable communication. In this respect, both scientific management and cybernetics, from their respective points of departure, can be seen as attempts to seriously think about the limits of communication in principle. For Wiener, at least, this seemed clear from the outset. Thus, much of his book *The Human Use of Human Beings*, according to him, concerned "the limits of communication within and among individuals."[7]

The more definitely the system draws the demarcation between communication and the incommunicable — in other words, defines the conditions for communication — the more rigorously it excludes what is external to it, which, however, is in itself a condition of all communication. This was absolutely clear in Taylorism, which excluded the multiplicity that did not fit the code of its functionality. The system was defined here as a machinery that does not take new ingredients from its environment; on the contrary, these were exactly what it aimed at eliminating. In the cybernetic system, however, the field of action has to be defined time and again, unlike in Taylorism where, once defined, it was regarded as remaining unchanged. It was characteristic of cybernetics to integrate the multiplicity into a common framework. It did not recognize any instance that would have been excluded in principle from communication, implying that ontological difference between communication and noncommunication, or the message and noise, could not be sustained as rigorously as in Taylorism. This is because it incorporated change and variety into the system. This was implicit in Shannon, and became the starting point of analysis in the second wave of cybernetics. Therefore, whereas scientific management had to anticipate in one solution all the contingencies it could face, the feedback-controlled system did precisely the opposite; it exploited contingencies as information that directed its behavior within certain pre-given limits. Although in the latter system also, it is true, the "state" or "internal unity" had to remain stable in order to enable continuous communication, the idea of a system differed: whereas for a mechanical system it was a perfect and unchanging machine, for a cybernetic one it was a continuously self-correcting one.[8]

This perspective was a precondition to thinking about these limits as an ontological structure for all communication, and thus to generalizing it as the condition of existence of communication as such. But, of course, these conditions are inarticulable, because if they were articulable, communication would not exist,

6 For the claim for universality in cybernetics, see an insightful paper by Geof Bowker, "How to be Universal: Some Cybernetic Strategies, 1943–70," *Social Studies of Science* 23, (1), (1993): 107–27.

7 Norbert Wiener, *The Human Use of Human Beings: Cybernetics and Society* (New York: Da Capo Press, 1954), 17.

8 Cybernetic systems can only adapt to external phenomena that they have been preprogrammed to recognize, and which must therefore be anticipated. Yet it remains true that for a cybernetic system, in contrast to a mechanical one, external phenomena — noise — is a source of information that directs the system's behavior. Therefore "anticipation" here has a somewhat different meaning; as for the former system what is anticipated had to be defined over and over again, whereas for the latter system this was done once and for all.

for it would be self-enclosed in the stillness of its unbroken sameness. On the other hand, what was going on in these communication projects was crystallized in the attempt to formalize communication. Not only were these types of projects endeavors to think of a permanent "essence" as an unchanging condition of historical communication but also they were manifestations of a cultural sentiment that was encapsulated in the ideals of a centralized order, a consistent meaning and a closed whole. They expressed an aspiration to understand the "sense" of a community as something that can be presented in terms of a general law. Yet, insofar as it is impossible to articulate the conditions of communication, it is equally impossible to formulate the principle of this communality as a law.

This does not, however, deprive these kinds of communication projects of the meaning they have as testimonies to the need to think about our own present time and to conceptualize related problems in a way that speaks directly to us. As marks of the present, they open up a historical relationship to the absolute basis for communication, although this opening is not entirely dependent on subjective decision. In fact, one can say more than that. Although we cannot know the absolute basis for communication, or the enabling condition in its difference from what is, we nevertheless inevitably have to make references to it. Even if thought cannot conceptualize it, making this difference between what is and its preconditions is necessary to communicative existence. As Walter Benjamin held, philosophy actually anticipates it in the existent.[9]

In the history of control of and by communication, "communication" was interpreted in terms of efficacy and legitimacy, for it was impossible to separate it from those instances that promoted and implemented social efficacy. As we move from an instrumentality of this kind to the constitutive dimension of communication, it becomes clear not only that communication produces effectiveness but also that the latter, combined with the scientific-technological rationalization of society, becomes increasingly a general social category, within which communication takes place and gets its meaningfulness. Yet, this category of effectiveness does not constitute any system of coordinates or unitary frame of reference for communication, for it is the latter that gives rise to these measurement systems and to their social establishment. It is also one actual "mechanism" for the rationalization of society: communication paves the way for it and marks its course, but this course is nothing other than communication itself, the rationalization of this communication.

If communication was often conceived of through a model inspired by the engineering and natural sciences, this is not to say that historical thought about communication needs to have been constantly in decline. First, what we have done here is to expound on a history of only one thread of communication. The presentation has followed a lineage in which communication has been rendered an ahistorical and instrumental object, but it goes without saying that communication is not reducible exclusively to this single history. As it is constituted historically

9 See Rodolphe Gasché, *Of Minimal Things: Studies on the Notion of Relation* (Stanford: Stanford University Press, 1999), 80.

in and through the process of its being articulated, communication cannot be thought of as a single, unified, and continuous entity, the history of which could be placed in a preexisting and unchanging framework with respect to the everyday realization of communicative relationships. A line of the thinking about communication can never be coherent, let alone linear. The path of development highlighted here is rather a sequence of interrelated but discontinuous formations and constellations of thought and practices. They are united by like style, manner and purpose, and the same historical destiny that is also our destiny, and it is precisely due to this community of fate that this path of development becomes significant for us.

Second, and related to the previous point, the processes of increasing technicization of communication cannot be conceived of exclusively in terms of a value judgment, that is normatively, because what they express and at the same time constitute is the historical context in which thinking about and articulation of communication can take place and in which it can assume its meaningfulness. As concepts of communication open up and establish a relationship to conditions for the social and give birth to temporary and discontinuous forms by way of which this relationship can be thought of and be exploited, they necessarily express the historicity of the social that is only conceivable within the realm of communication. Attempts to conceptualize communication are therefore always at the same time endeavors to think about the conditions for community. And, in the process of this thinking, they constitute a historical domain or perspective within which thinking about community is possible. This is why communication is nothing other than the unfolding of the social and thought about this historical unfolding.

Therefore, it is not a question of the scope of thought about communication being reduced or made ahistorical as such by Taylor, Wiener, Shannon, and a swarm of other less well-known figures. Rather, what happened to communication in Taylorism or in cybernetics was, to a certain extent, immanent to communication itself. Clearly the formalization of communication was a consequence of at least three related and concurrent developments, all developments of and in communication. These are the increase in number and complexity of processes of communication, the development of methods, techniques, and structures of communication, and the professionalization of scientific research. Together they all intensified the study of communication, with the courses, channels, and tracks of communication becoming an object of persevering analysis and theory formation. If this formalization was the precondition of telecommunication, the formalized units and relationships are, essentially, communication's own units and relationships. Seen from the point of view of this study, the formalization of communication is inevitable because of our practical needs. But then again, this formalization should not be regarded as something depending on a personal decision. Rather, it is originally part of the modernization of communication, part of its unfolding and appearance regardless of any single decision or act. Through the process of formalization we can recognize the history of modernity as our history, and this experience of recognition is itself nothing other than an experience of and in communication.

Bibliography

Official Publications of the United States

Annual Report of the Postmaster General: The Fiscal Year Ended June 30, 1871. Washington, DC: Government Printing Office.

Annual Report of the Postmaster General: The Fiscal Year Ended June 30, 1873. Washington, DC: Government Printing Office.

Federal Communications Commission. *Investigation of the Telephone Industry in the United States*. New York: Arno Press. Reprint of House Doc. No. 340, 76th Cong., 1st Sess. Washington, DC: Government Printing Office, 1974 [1939].

Congressional Globe. 34th Cong., 3d Sess. (1857). Washington, DC: John C. Rives.

Congressional Globe. 36th Cong., 1st Sess. (1860). Washington, DC: John C. Rives.

Congressional Record. 48th Cong., 1st Sess., vol. 15 (1884). Washington, DC: Government Printing Office.

President's Communications Policy Board. *Telecommunications: A Program for Progress; A Report by the President's Communications Policy Board*. In *Documents in American Telecommunications Policy: Volume Two*, edited by John Kittross. New York: Arno Press, 1977 [1951].

United States Congress. House Doc. No. 156, 19th Cong., 1st Sess. (1826).

United States Congress. House Doc. No. 340, 76th Cong., 1st Sess. (1939).

United States Congress. House Report No. 187, 28th Cong., 2d Sess. (1845).

United States Congress. House Ex. Doc. No. 5, 31st Cong., 1st Sess. (1849).

United States Congress. Senate Ex. Doc. No. 1, 31st Cong., 2d Sess. (1850).

United States Congress. House Ex. Doc. No. 1, Part 3, 41st Cong., 2d Sess. (1869).

United States Congress. Senate Doc. No. 399, 63d Cong., 2d Sess. (1914). Washington, DC: Government Printing Office.

Periodicals

American Telegraph Magazine
Andover Review
Atlantic Monthly
Bell System Technical Journal
Bell Telephone Quarterly
Boston Daily Advertiser
Bulletin of American Geographical and Statistical Society
Democratic Review
Electrical Communication
Harper's Magazine
Harper's Weekly
Independent
Journal of Policy History
Journal of Politics
McClure's Magazine
North American Review
Political Science Quarterly
Post Office Electrical Engineers' Journal
Princeton Review
Reader

Scientific American
Telecommunications Policy
Telegrapher
Telephone Engineer
Telephony
Yale Law Journal

Literature

Ahvenainen, Jorma. *The Far Eastern Telegraphs: The History of Telegraphic Communications between the Far East, Europe and America before the First World War*. Jyväskylä: Gummerus, 1981.
Akin, William. *Technocracy and the American Dream: The Technocrat Movement, 1900–1941*. Berkeley: University of California Press, 1977.
Amadae, S. M. *Rationalizing Capitalist Democracy: The Cold War Origins of Rational Choice Liberalism*. Chicago: University of Chicago Press, 2003.
"Application of the Modern Telephone." *Telephony* (August, 1902).
Archer, Charles. *Guide to the Electric Telegraph*. London: Smith, 1852.
Arkright, Frank. *The ABC of Technocracy*. London: Hamish Hamilton, 1933.
Arnold, H. D. "An Introduction to Research in the Communication Field." In *Modern Communication*, edited by Arthur Page, H. D. Arnold, John E. Otterson, Harvey Fletcher, Ralph Brown, Frank B. Jewett, and Herbert E. Ives. Boston: Houghton Mifflin, 1932.
Arnold, Thurman. *The Folklore of Capitalism*. Washington, DC: Beard Books, 2000 [1937].
Aronson, Sidney. "Bell's Electrical Toy: What's the Use? The Sociology of Early Telephone Usage." In *The Social Impact of the Telephone*, edited by Ithiel de Sola Pool. Cambridge: MIT Press, 1977.
Arrighi, Giovanni. *The Long Twentieth Century: Money, Power, and the Origins of Our Times*. London: Verso, 1994.
Arrow, Kenneth. *Social Choice and Individual Values*. New York: Wiley, 1951.
"Associated Press Service by Telephone." *Telephone Engineer* 5, (5), (April 1911).
"The Associated Press: What It Is!" *Telegraph and Telephone Age* (January 16, 1916).
Atlan, Henri. "On a Formal Definition of Organization." *Journal of Theoretical Biology* 45, (1974).
Ault, Phil. *Wires West*. New York: Dodd, Mead & Co., 1974.
Banta, Martha. *Taylored Lives: Narrative Productions in the Age of Taylor, Veblen, and Ford*. Chicago: University of Chicago Press, 1993.
Barnard, Chester. *The Functions of the Executive* (30th anniversary edn). Cambridge, MA: Harvard University Press, 1968 [1938].
Barrett, R. T. "The Growth of America's Communication Needs." *Bell Telephone Quarterly* 12, (4), (1933).
Barrett, Robertson. "The Telephone as a Social Force." *Bell Telephone Quarterly* 19, (2), (1940).
Barry, Andrew. "Lines of Communication and Spaces of Rule." In *Foucault and Political Reason: Liberalism, Neo-Liberalism and Rationalities of Government*, edited by Andrew Barry, Thomas Osborne, and Nikolas Rose. London: UCL Press, 1996.
Bates, David. *Lincoln in the Telegraph Office: Recollections of the Military Telegraph Corps during the Civil War*. Introduction by James Rawley. Lincoln: University of Nebraska Press, 1995 [1907].
Beard, Charles and Beard, William. *The American Leviathan: The Republic in the Machine Age*. London: Jonathan Cape, 1930.
Beer, Stafford. *Cybernetics and Management*. London: English Universities Press, 1959.
— *Brain of the Firm: The Managerial Cybernetics of Organization*. London: Allen Lane, Penguin Press, 1972.
Beetham, David. *Bureucracy* (2nd edn). Buckingham: Open University Press, 1996.
Bell, Daniel. *The End of Ideology: On The Exhaustion of Political Ideas in the Fifties*. Cambridge, MA: Harvard University Press, 1960.
— Preface to *The Age of Automation*, by Sir Leon Bagrit. New York: Mentor, 1965.
— "The Social Framework of the Information Society." In *The Microelectronics Revolution*, edited by Tom Forester. Oxford: Basil Blackwell, 1980.
Beniger, James. *The Control Revolution: Technological and Economic Origins of the Information Society*. Cambridge, MA: Harvard University Press, 1986.
— "Conceptualizing Information Technology as Organization, and Vice Versa." In *Organizations and Communication Technology*, edited by Janet Fulk and Charles Steinfield. Newbury Park: Sage, 1990.
Benjamin, A. S. "Teletype: The Telephone Typewriter." *Telephone Engineer* 33, (12), (December 1929).

Bentley, Arthur. *The Process of Government: A Study of Social Pressures*. Evanston: Principia Press of Illinois, 1935 [1908].
Bertalanffy, Lugwig von. *General System Theory: Foundations Development Applications*. New York: Braziller, 1968.
Bethell, Union. *Telephone Engineer* 9, (4), (1913).
Bickers, Kenneth. "Transformations in the Governance of the American Telecommunications Industry." In *Governance of the American Economy*, edited by John Campbell, J. Rogers Hollingsworth, and Leon Lindberg. Cambridge: Cambridge University Press, 1991.
Bierstedt, Robert. *American Sociological Theory: A Critical History*. New York: Academic Press, 1981.
Bimber, Bruce. *Information and American Democracy: Technology in the Evolution of Political Power*. Cambridge: Cambridge University Press, 2003.
Blain, Ray. "Communication is an Element in National Progress." *Telephone Engineer* 33, (3), (March, 1929).
Blondheim, Menahem. *News over the Wires: The Telegraph and the Flow of Public Information in America, 1844–1897*. Cambridge, MA: Harvard University Press, 1994.
Boguslaw, Robert. *The New Utopians: A Study of System Design and Social Change*. Englewood Cliffs: Prentice-Hall, 1965.
Bohn, Frank. "The Association and Its Needs." *Telephone Engineer* 24, (4), (October 1920).
Boland, R. J., Jr., and Hirschheim, R. A., eds. *Critical Issues in Information Systems Research*. Chichester: John Wiley & Sons, 1987.
Boorstin, Daniel. *The Americans: The National Experience*. London: Sphere Books, 1988 [1965].
Boulding, Kenneth. *The Organizational Revolution: A Study in the Ethics of Economic Organization*. New York: Harper & Brothers, 1953.
— "General Systems Theory: The Skeleton of Science." In *General Systems Theory and Human Communication*, edited by Brent Ruben and John Kim. Rochelle Park: Hayden Book Co., 1975 [1956].
Bowker, Geof. "How to be Universal: Some Cybernetic Strategies, 1943–70." *Social Studies of Science* 23, (1), (1993): 107–27.
Bowman, Scott. *The Modern Corporation and American Political Thought: Law, Power, and Ideology*. University Park: Pennsylvania State University Press, 1996.
Braverman, H. *Labor and Monopoly Capital: The Degradation of the Work in the Twentieth Century*. New York: Monthly Review Press, 1974.
Braymer, D. H. "New Police Telephone and Signal System." *Telephone Engineer* 3, (3), (March 1910).
Briggs, Asa and Burke, Peter. *A Social History of the Media: From Gutenberg to the Internet*. Cambridge: Polity Press, 2002.
Briggs, Charles and Maverick, Augustus. *The Story of the Telegraph, and a History of the Great Atlantic Cable*. New York: Rudd & Carleton, 1858.
Brinkley, Alan. *Liberalism and Its Discontents*. Cambridge, MA: Harvard University Press, 1998.
Brock, Gerald. *The Telecommunications Industry: The Dynamics of Market Structure*. Cambridge, MA: Harvard University Press, 1981.
— *Telecommunication Policy for the Information Age: From Monopoly to Competition*. Cambridge, MA: Harvard University Press, 1994.
Brooks, John. *Telephone: The First Hundred Years*. New York: Harper & Row, 1975.
Brooks, Sydney. "Politics of American Business." *North American Review* 93, (May 1911).
Brown, Richard. *Knowledge is Power: The Diffusion of Information in Early America, 1700–1865*. New York: Oxford University Press, 1989.
Burchell, Graham, Gordon, Colin, and Miller, Peter, eds. *The Foucault Effect: Studies in Governmentality*. Exeter: Harvester Wheatsheaf, 1991.
Burlingame, Roger. *March of the Iron Men: A Social History of Union through Invention*. New York: Charles Scribner's Sons, 1939.
— *Engines of Democracy: Inventions and Society in Mature America*. New York: Charles Scribner's Sons, 1940.
Bush, Vannevar. *Science: The Endless Frontier*. Washington, DC: Government Printing Office, 1945.
Campagna, Anthony. *U.S. National Economic Policy 1917–1985*. New York: Praeger, 1987.
Campbell, Ballard. *The Growth of American Government: Governance from the Cleveland Era to the Present*. Bloomington: Indiana University Press, 1995.
Campbell, James. *Understanding John Dewey: Nature and Cooperative Intelligence*. Chicago: Open Court, 1995.
Cannon, Walter. *The Wisdom of the Body* (revised edn). New York: W. W. Norton & Co., 1939 [1932].
Carey, James. "A Cultural Approach to Communication." *Communication* 2, (1975): 1–22.

— "Technology and Ideology: The Case of the Telegraph." In *Communication as Culture: Essays on Media and Society*, edited by James Carey. London: Unwin Hyman, 1989.
— "Time, Space, and the Telegraph." In *Communication in History: Technology, Culture, Society*, edited by David Crowley and Paul Heyer. New York: Longman, 1991.
Carlson, W. Bernard. "The Telephone as Political Instrument: Gardiner Hubbard and the Formation of the Middle Class in America, 1875–1880." In *Technologies of Power: Essays in Honor of Thomas Parke Hughes and Agatha Chipley Hughes*, edited by Michael Allen and Gabrielle Hecht. Cambridge: MIT Press, 2001.
Carty, John. "Ideals of the Engineer." *Bell Telephone Quarterly* 8, (2), (1928).
Cass, Ronald. "Titles IV, V, and VI: Review, Enforcement, and Power under the Communications Act of 1934; Choice and Change in Institutional Design." In *A Legislative History of the Communications Act of 1934*, edited by Max Paglin. New York: Oxford University Press, 1989.
Casson, Herbert. *The History of the Telephone*. Chicago: A. C. McClurg & Co., 1910.
— "The Telephone as it is To-Day." *World's Work* 19, (April 1910).
— "The Future of the Telephone." *World's Work* 19, (May 1910).
— "The Social Value of the Telephone." *Independent*, October 26, 1911.
Chambers Dictionary of Etymology. Edited by Robert K. Barnhart. Edinburgh: Chambers, 1988.
Chandler, Alfred, Jr. *The Visible Hand: The Managerial Revolution in American Business*. Cambridge, MA: Belknap Press of Harvard University Press, 1977.
Channing, William. "Morse's Telegraph for Fire Alarms." *Boston Daily Advertiser and Patriot*, June 3, 1845.
— *Notes on the Medical Applications of Electricity*. Boston: Daniel Davis Jr., 1849.
— "Report Concerning a Uniform System of Fire Alarms." *City Document No. 42, City of Boston*, March 24, 1851.
— "On the Municipal Electric Telegraph: Especially in Its Application to Fire Alarms." *American Journal of Science* 13, (1852): 58–83.
— "The American Fire-Alarm Telegraph." *Ninth Annual Report of the Smithsonian Institution* (1855): 147–55.
Chapuis, Robert. *100 Years of Telephone Switching (1878–1978): Part I; Manual and Electromechanical Switching (1878–1960s)*. Amsterdam: North-Holland Publishing Co., 1982.
Checkland, Peter. "Systems Theory and Management Thinking." *American Behavioral Scientist* 38, (1), (1994): 75–91.
— *Systems Thinking, Systems Practice*. Chichester: John Wiley & Sons, 1999.
Cherry, Colin. "The Telephone System: Creator of Mobility and Social Change." In *The Social Impact of the Telephone*, edited by Ithiel de Sola Pool. Cambridge: MIT Press, 1977.
— *On Human Communication* (3rd edn). Cambridge: MIT Press, 1980 [1957].
Childs, Frank. "When the Telephone was Young in Washington, D.C." *Telephony* 104, (13), (April 1, 1933).
Churchman, C. West, Ackoff, Russell, and Arnoff, Leonard. *Introduction to Operations Research*. New York: John Wiley & Sons, 1957.
Clapp, W. H. "Telegraph or Telephone for Railroad Work?" *Telephone Engineer* 11, (4), (April 1914).
Clark, Christopher. "The Consequences of the Market Revolution in the American North." In *The Market Revolution in America: Social, Political, and Religious Expressions, 1800–1880*, edited by Melvyn Stokes and Stephen Conway. Charlottesville: University Press of Virginia, 1996.
Cmiel, Kenneth. "On Cynicism, Evil, and the Discovery of Communication in the 1940s." *Journal of Communication* 46, (3), (Summer 1996): 88–107.
Coar, C. H. "Encouraging Efficient Employees." *Telephone Engineer* 9, (4), (1913).
Coe, Lewis. *The Telegraph: A History of Morse's Invention and Its Predecessors in the United States*. Jefferson: McFarland & Co., 1993.
Cohen, Jeffrey. "The Telephone Problem and the Road to Telephone Regulation in the United States, 1876–1917." *Journal of Policy History* 3, (1), (1991): 42–69.
Colton, Arthur. "The Telephone Newspaper: New Experiment in America." *Telephony* 62, (13), (1912).
"Communicating Over Great Distances." *Scientific American* (June 5, 1915).
"Communications Report to Commerce Committee." *Telephone Engineer* 38, (2), (February 1934).
Conway, Flo and Siegelman, Jim. *Dark Hero of the Information Age: In Search of Norbert Wiener, the Father of Cybernetics*. New York: Basic Books, 2005.
Cooley, Charles. "The Theory of Transportation." *Publications of the American Economic Association* 9, (3), (1894).
— "The Process of Social Change." *Political Science Quarterly* 12, (1), (1897).
— *Human Nature and the Social Order*. New York: Scribner, 1902.

— *Social Organization: A Study of the Larger Mind.* New York: Charles Scribner's Sons, 1912.
— *Sociological Theory and Social Research.* New York: Henry Holt & Co, 1930.
— *On Self and Social Organization,* edited by Hans-Joachim Schubert. Chicago: University of Chicago Press, 1998.
Cooper, Dennis and Harris, E. L. *The People Machine: An Illustrated History of the Telephone on the Center West Coast of Florida* (3rd edn). Tampa: GTE Florida, 1988 [1971].
Cortada, James. *Strategic Data Processing: Considerations for Management.* Englewood Cliffs: Prentice-Hall, 1984.
Cowan, Ruth. *A Social History of American Technology.* New York: Oxford University Press, 1997.
Cox, Archibald. *The Court and the Constitution.* Boston: Houghton Mifflin, 1987.
Cox, Kenneth and Byrnes, William. "Title II: The Common Carrier Provisions; A Product of Evolutionary Development." In *A Legislative History of the Communications Act of 1934,* edited by Max Paglin. New York: Oxford University Press, 1989.
Croly, Herbert. *Progressive Democracy.* New York: Macmillan, 1915.
Curti, Merle. *The Roots of American Loyalty.* New York: Columbia University Press, 1946.
Czitrom, Daniel. *Media and the American Mind: From Morse to McLuhan.* Chapel Hill: University of North Carolina Press, 1982.
Dandeker, Christopher. *Surveillance, Power and Modernity: Bureaucracy and Discipline from 1700 to the Present Day.* Cambridge: Polity Press, 1990.
Danielian, N. R. *A.T.&T.: The Story of Industrial Conquest.* New York: Vanguard Press, 1939.
Davies, Andrew. *Telecommunications and Politics: The Decentralised Alternative.* London: Pinter Publishers, 1994.
Davis, Whitney. "Decadence and the Organic Metaphor." *Representations* 89, (2005): 131–49.
Dawley, Alan. *Struggles for Justice: Social Responsibility and the Liberal State.* Cambridge, MA: Belknap Press of Harvard University Press, 1991.
Dean, Mitchell. "Foucault, Government and the Enfolding of Authority." In *Foucault and Political Reason: Liberalism, Neo-Liberalism and Rationalities of Government,* edited by Andrew Barry, Thomas Osborne, and Nikolas Rose. London: UCL Press, 1996.
— *Governmentality: Power and Rule in Modern Society.* London: Sage, 1999.
Debray, Régis. *Media Manifestos: On the Technological Transmission of Cultural Forms.* Translated by Eric Rauth. New York: Verso, 1996.
— *Transmitting Culture.* Translated by Eric Rauth. New York: Columbia University Press, 2000.
Dechert, Charles. "The Development of Cybernetics." In *The Social Impact of Cybernetics,* edited by Charles Dechert. Notre Dame: University of Notre Dame, 1966.
Deleuze, Gilles. *The Logic of Sense.* Translated by Mark Lester with Charles Stivale. Edited by Constantin Boundas. London: Athlone Press, 1990.
Denison, Merrill. *Advancing America: The Drama of Transportation and Communication.* New York: Dodd, Mead & Co., 1936.
Dennard, Linda. "Neo-Darwinism and Simon's Bureaucratic Antihero." *Administration & Society* 26, (4), (1995): 464–87.
Deutsch, Karl. "Mechanism, Organism, and Society: Some Models in Natural and Social Science." *Philosophy of Science* 18, (3), (1951): 230–52.
— "Mechanism, Teleology, and Mind." *Philosophy and Phenomenological Research* 12, (2), (1951): 185–223.
— *The Nerves of Government.* Glencoe, IL: Free Press, 1966.
Dewey, John. "Ethics and Physical Science." *Andover Review* 7, (42), (1887).
— "The School as Social Center." *Proceedings of the National Education Association* (1902).
— *Democracy and Education: An Introduction to the Philosophy of Education.* New York: Macmillan, 1917.
Dewey, John. *The Public and Its Problems: An Essay in Political Inquiry.* Chicago: Gateway Books, 1946 [1927].
Diebold, John. *Automation: The Advent of the Automatic Factory.* New York: van Nostrand, 1952.
Dilts, Marion. *The Telephone in a Changing World.* New York: Longmans, Green & Co., 1941.
"A Discussion on Railway Telephony." *Telephone Engineer* 12, (4), (October, 1914).
Donham, Wallace. *Business Looks at the Unforeseen.* New York: Whittlesey House, McGraw-Hill Book Co., 1932.
Doray, Bernard. *From Taylorism to Fordism: A Rational Madness.* Translated by David Macey. London: Free Association Books, 1988.
Draper, John. *Thoughts on the Future Civil Policy of America.* New York: Harper & Brothers, 1865.

Du Boff, Richard. "The Telegraph and the Structure of Markets in the United States: 1845–1890." *Research in Economic History* 8, (1983).

— "The Telegraph in Nineteenth-Century America: Technology and Monopoly." *Comparative Studies in Society and History* 26 (1984): 571–86.

Dudley, Leonard. *The Word and the Sword: How Techniques of Information and Violence have Shaped our World.* Ann Arbor, MI: Blackwell, 1991.

Duncan, Hugh. *Communication and Social Order.* New York: Bedminster Press, 1962.

Dwyer, John. *To Wire the World: Perry M. Collins and the North Pacific Telegraph Expedition.* Westport, CT: Praeger, 2001.

Easton, David. *The Political System: An Inquiry into the State of Political Science.* New York: Knopf, 1953.

— *A Framework for Political Analysis.* Englewood Cliffs: Prentice-Hall, 1965.

Edwards, Paul. *The Closed World: Computers and the Politics of Discourse in Cold War America.* Cambridge: MIT Press, 1996.

"Efficiency Engineering in the Telegraph Service." *Telegraph and Telephone Age* (January 1, 1916).

Eisenach, Eldon. *The Lost Promise of Progressivism.* Lawrence: University Press of Kansas, 1994.

Elsner, Henry, Jr. *The Technocrats: Prophets of Automation.* Syracuse: Syracuse University Press, 1967.

Endres, K. W. "The Telephone for Train Dispatching." *Telephone Engineer* 3, (4), (April 1910).

Eriksson, Kai. "On the Ontology of Networks." *Communication and Critical/Cultural Studies* 2, (4), (2005): 305–23.

— "Foucault, Deleuze, and the Ontology of Networks." *European Legacy* 10, (6), (2005): 595–610.

Everitt, William. "Telecommunications: The Resource Not Depleted by Use; A Historical and Philosophical Resumé." *Proceedings of the IEEE*, 64, (9), (1976).

Fagen, M., ed. *A History of Engineering and Science in the Bell System: 1. The Early Years, 1875–1925.* New York: Bell Telephone Laboratories, 1975.

Farrah, Barbara and Maxwell, Mike. "Building America's Infostructure: Public Policy in the Information Age." *Telephony* (April 20, 1992).

Faulkner, Harold. *American Economic History* (8th edn). New York: Harper & Row, 1964.

Fawcett, Waldon. "How Uncle Sam Uses the Telephone." *Telephony* 19, (4), (January 1910).

— "Telephones at United States Forts." *Telephone Engineer* 5, (4), (April 1911).

Field, Alexander. "The Magnetic Telegraph, Price and Quantity Data, and the New Management of Capital." *Journal of Economic History* 52, (2), (1992): 401–13.

Fine, Sidney. *Laissez Faire and the General-Welfare State: A Study of Conflict in American Thought, 1865–1901.* Ann Arbor, MI: University of Michigan Press, 1956.

Fischer, Claude. *America Calling: A Social History of the Telephone to 1940.* Berkeley: University of California Press, 1992.

Fischer, Frank. *Technocracy and the Politics of Expertise.* London: Sage, 1990.

Flichy, Patrice. *Dynamics of Modern Communication: The Shaping and Impact of New Communication Technologies.* Translated by Liz Libbrecht. London: Sage, 1995.

Flood, Robert. *Rethinking The Fifth Discipline: Learning within the Unknowable.* London: Routledge, 1999.

Flueckiger, Gerald. *Control, Information, and Technological Change.* Dordrecht: Kluwer Academic Publishers, 1995.

Fones-Wolf, Elizabeth. *Selling Free Enterprise: The Business Assault on Labor and Liberalism, 1945–60.* Urbana: University of Illinois Press, 1994.

Forrester, Jay. *Industrial Dynamics.* Cambridge: MIT Press, 1968 [1961].

Fortun, M. and Schweber, S. S. "Scientists and the Legacy of World War II: The Case of Operations Research (OR)." *Social Studies of Science* 23, (4), (1993): 595–642.

Fott, David. *John Dewey: America's Philosopher of Democracy.* Lanham: Rowman & Littlefield, 1998.

Foucault, Michel. "Governmentality." In *The Foucault Effect: Studies in Governmentality*, edited by Graham Burchell, Colin Gordon, and Peter Miller. Hemel Hempstead: Harvester, 1991.

Fuhrman, Ellsworth. *The Sociology of Knowledge in America, 1883–1915.* Charlottesville: University Press of Virginia, 1980.

Fuller, Wayne. *The American Mail: Enlarger of the Common Life.* Chicago: University of Chicago Press, 1972.

Furnas, J. C. *The Americans: A Social History of the United States, 1587–1914.* New York: G. P. Putnam's Sons, 1969.

Gabler, Edwin. *The American Telegrapher: A Social History, 1860–1900.* New Brunswick: Rutgers University Press, 1988.

Galambos, Louis and Pratt, Joseph. *The Rise of the Corporate Commonwealth: U.S. Business and Public Policy in the Twentieth Century.* New York: Basic Books, 1988.

Galison, Peter. "The Ontology of the Enemy: Norbert Wiener and the Cybernetic Vision." *Critical Inquiry* 21, (1), (1994): 228–66.

Garraty, John. *The New Commonwealth, 1877–1890*. New York: Harper & Row, 1968.

Garry, Patrick. *Liberalism and American Identity*. Kent: Kent State University Press, 1992.

Gasché, Rodolphe. *Of Minimal Things: Studies on the Notion of Relation*. Stanford: Stanford University Press, 1999.

George, F. H. *Automation, Cybernetics and Society*. London: Leonard Hill, 1959.

Gettell, Raymond. *History of American Political Thought*. New York: Century, 1928.

Gherardi, Bancroft and Jewett, F. B. "Telephone Communication System of the United States." *Bell System Technical Journal* 23, (9), (January 1930).

Gillman, Howard. *The Constitution Besieged: The Rise and Demise of Lochner Era Police Powers Jurisprudence*. Durham: Duke University Press, 1993.

Glanville, Ranulph. "A Cybernetic Musing: In the Animal and the Machine." *Cybernetics and Human Knowing* 4, (4), (1997).

Goodnow, Frank. *Politics and Administration: A Study in Government*. New York: Macmillan, 1900.

Gordon, Colin. *New Deals: Business, Labor, and Politics in America, 1920–1935*. Cambridge: Cambridge University Press, 1994.

Gordon, John. *A Thread Across the Ocean: The Heroic Story of the Transatlantic Cable*. New York: Walker & Co., 2002.

Goulden, Joseph. *Monopoly*. New York: G. P. Putnam's Sons, 1968.

"Government Stations at Washington." *Telephone Engineer* 5, (6), (June 1911).

Grabbe, E. M. "The Language of Automation," in *Automation in Business and Industry*, edited by E. M. Grabbe. New York: John Wiley & Sons, 1957.

Grace, S. P. "Common Battery for Small Exchanges." *Telephony* 4, (2), (1902).

"Growth of the Telephone," *Wall Street Journal* (June 17, 1914).

Haar, Michel. *The Song of the Earth: Heidegger and the Grounds of the History of Being*. Translated by Reginald Lilly. Bloomington: Indiana University Press, 1993.

Haber, Samuel. *Efficiency and Uplift: Scientific Management in the Progressive Era, 1890–1920*. Chicago: University of Chicago Press, 1964.

Hall, Peter. *The Organization of American Culture, 1700–1900: Private Institutions, Elites, and the Origins of American Nationality*. New York: New York University Press, 1982.

Hamilton, Alexander, Madison, James and Jay, John. *The Federalist Papers*. New York: New American Library, 1961.

Hanson, Russell. *The Democratic Imagination in America: Conversations with Our Past*. Princeton: Princeton University Press, 1985.

Haraway, Donna. *Simians, Cyborgs, and Women: The Reinvention of Nature*. New York: Routledge, 1991.

Hardt, Hanno. *Social Theories of the Press: Early German & American Perspectives*. Beverly Hills: Sage, 1979.

— *Critical Communication Studies: Communication, History and Theory in America*. London: Routledge, 1992.

Harlow, Alwin. *Old Wires and New Waves: The History of the Telegraph, Telephone, and Wireless*. New York: D. Appleton-Century Co., 1936.

Harp, Gillis. *Positivist Republic: Auguste Comte and the Reconstruction of American Liberalism, 1865–1920*. University Park: Pennsylvania State University Press, 1995.

Hart, David. *Forged Consensus: Science, Technology, and Economic Policy in the United States, 1921–1953*. Princeton: Princeton University Press, 1998.

Hartley, Ralph. "Transmission of Information." *Bell System Technical Journal* 7, (July 1928): 535–63.

Hartz, Louis. *The Liberal Tradition in America: An Interpretation of American Political Thought since the Revolution*. New York: Harcourt, Brace & World, 1955.

Haskell, Thomas. *The Emergence of Professional Social Science: The American Social Science Association and the Nineteenth-Century Crisis of Authority*. Urbana: University of Illinois Press, 1977.

Hayles, N. Katherine. *How We Became Posthuman: Virtual Bodies in Cybernetics, Literature, and Informatics*. Chicago: University of Chicago Press, 1999.

Headrick, Daniel. *The Invisible Weapon: Telecommunications and International Politics, 1851–1945*. New York: Oxford University Press, 1991.

— "Public–Private Relations in International Telecommunications before World War II." In *Telecommunications Politics: Ownership and Control of the Information Highway in Developing Countries*, edited by Bella Mody, Johannes M. Bauer, and Joseph D. Straubhaar. Mahwah: Lawrence Erlbaum Associates, 1995.

Heidegger, Martin. *Antwort: Martin Heidegger im Gespräch*, edited by Günther Neske and Emil Kettering. Pfullinger: Neske, 1988.
Heims, Steve. "Encounter of Behavioral Sciences with New Machine-Organism Analogies in the 1940's." *Journal of the History of the Behavioral Sciences* 11, (4), (October 1975): 368–73.
— *John von Neumann and Norbert Wiener: From Mathematics to the Technologies of Life and Death*. Cambridge: MIT Press, 1980.
— *The Cybernetics Group*. Cambridge: MIT Press, 1991.
Henderson, C. R. *The Social Spirit in America*. New York: Chautauqua-Century Press, 1897.
Hendrick, Burton. "Telephones for the Millions." *McClure's Magazine* 44, (1914).
Hibbard, Angus. "Telephones in Central Station Control." *Telephone Engineer* 10, (1), (July 1913).
Hickman, Larry. *John Dewey's Pragmatic Technology*. Bloomington: Indiana University Press, 1990.
— *Philosophical Tools for Technological Culture: Putting Pragmatism to Work*. Bloomington: Indiana University Press, 2001.
Highton, Edward. *The Electric Telegraph: Its History and Progress*. London: John Weale, 1852.
Hilton, Alice. Foreword to *The Evolving Society: The Proceedings of the First Annual Conference on the Cybercultural Revolution; Cybernetics and Automation*, edited by Alice Hilton. New York: Institute for Cybercultural Research, 1966.
Hirschhorn, Larry. *Beyond Mechanization: Work and Technology in a Postindustrial Age*. Cambridge: MIT Press, 1984.
"The Historical Search for Significance." In *The Significance of the Media in American History*, edited by James Startt and W. David Sloan. Northport: Vision Press, 1994.
Hitch, Charles. "A Planning-Programming-Budgeting System." In *Science, Technology, and Management*, edited by Fremont Kast and James Rosenzweig. New York: McGraw-Hill Book Co., 1963.
Hochheimer, Lewis. *The Law of Crimes and Criminal Procedure* (2nd edn). Baltimore: Baltimore Book Co., 1904.
Hofstadter, Richard. *Social Darwinism in American Thought*. New York: George Braziller, 1969.
Horwitz, Morton. *The Transformation of American Law, 1780–1860*. Cambridge, MA: Harvard University Press, 1977.
Horwitz, Robert. *The Irony of Regulatory Reform: The Deregulation of American Telecommunications*. New York: Oxford University Press, 1989.
Hounshell, David. *From the American System to Mass Production, 1800–1932: The Development of Manufacturing Technology in the United States*. Baltimore: Johns Hopkins University Press, 1984.
House, Richard. "The Encyclopedia Complex: Contemporary Narratives of Information." *SubStance* 92, (2), (2000): 25–46.
Hubbard, Gardiner. "The Proposed Changes in the Telegraphic System." *North American Review* 117, (240), (1873): 80–107.
Huber, Peter, Kellogg, Michael, and Thorne, John. *Federal Telecommunications Law*. Boston: Little, Brown & Co., 1992.
Hughes, Thomas. *Rescuing Prometheus*. New York: Vintage Books, 1998.
Hugill, Peter. *Global Communications since 1844: Geopolitics and Technology*. Baltimore: Johns Hopkins University Press. 1999.
Hunter, Howard. "The Teletypewriter Helps the Police." *Bell Telephone Quarterly* 18, (3), (1939).
Hurst, James. *Law and Markets in United States History: Different Modes of Bargaining among Interests*. Madison: University of Wisconsin Press, 1982.
Hyman, Harold and Wiecek, William. *Equal Justice under Law: Constitutional Development, 1835–1875*. New York: Harper & Row, 1982.
"The Indispensable Telephone." *Telephony* 12, (5), (November 1906).
"Influence of the Telegraph upon Literature." *Democratic Review* 22, (May 1848).
Innis, Harold. *The Press: A Neglected Factor in the Economic History of the Twentieth Century*. London: Oxford University Press, 1949.
"The Intellectual Effects of Electricity." *Spectator* 63 (November 9, 1889).
Introna, L. D. "Management Information Systems." In *Encyclopedia of Communication and Information*, edited by J. R. Schement. New York: Macmillan Reference, USA, 2001.
Irwin, Manley. *Telecommunications America: Markets without Boundaries*. Westport, CT: Quorum Books, 1984.
Israel, Paul. *From Machine Shop to Industrial Laboratory: Telegraph and the Changing Context of American Invention, 1830–1920*. Baltimore: Johns Hopkins University Press, 1992.
Jewett, Frank. "The Telephone Switchboard: Fifty Years of History." *Bell Telephone Quarterly* 8, (1928).
— "The Social Implications of Scientific Research in Electrical Communication." *Bell Telephone Quarterly* 15, (4), (1936).

— "A Quarter Century of Transcontinental Telephony." In *The Telephone: An Historical Anthology*, edited by George Shiers. New York: Arno Press, 1977 [1940].

John, Richard. "American Historians and the Concept of the Communications Revolution." In *Information Acumen: The Understanding and Use of Knowledge in Modern Business*, edited by Lisa Bud-Frierman. London: Routledge, 1994.

— *Spreading the News: The American Postal System from Franklin to Morse.* Cambridge, MA: Harvard University Press, 1995.

Jones, Alexander. *Historical Sketch of the Electric Telegraph: Including Its Rise and Progress in the United States.* New York: George P. Putnam, 1852.

Jordan, John. *Machine-Age Ideology: Social Engineering and American Liberalism, 1911–1939.* Chapel Hill: University of North Carolina Press, 1994.

Judson, Katharine, ed. *Selected Articles on Government Ownership of Telegraph and Telephone.* White Plains: H. W. Wilson, 1914.

Kahn, Paul. *Legitimacy and History: Self-Government in American Constitutional Theory.* New Haven: Yale University Press, 1992.

Kanigel, Robert. *The One Best Way: Frederick Winslow Taylor and the Enigma of Efficiency.* London: Abacus, 1997.

Karl, Barry. *Executive Reorganization and Reform in the New Deal: The Genesis of Administrative Management, 1900–1939.* Cambridge, MA: Harvard University Press, 1963.

Kasson, John. *Civilizing the Machine: Technology and Republican Values in America, 1776–1900.* Harmondsworth: Penguin Books, 1977.

Kast, Fremont and Rosenzweig, James, eds. *Science, Technology, and Management.* New York: McGraw-Hill Book Co., 1963.

Katz, Elihu and Lazarsfeld, Paul F. *Personal Influence.* New York: Free Press, 1955.

Katz, James. *Connections: Social and Cultural Studies of the Telephone in American Life.* New Brunswick: Transaction Publishers, 1999.

Kaufman-Osborn, Timothy. "John Dewey and the Liberal Science of Community." *Journal of Politics* 46, (4), (1984): 1142–65.

Keller, Morton. *Affairs of State: Public Life in Late Nineteenth Century America.* Cambridge, MA: Belknap Press of Harvard University Press, 1977.

— "The Pluralistic State: American Economic Regulation in Comparative Perspective, 1900–1930." In *Regulation in Perspective*, edited by Thomas McCraw. Cambridge, MA: Harvard University Press, 1981.

— *Regulating a New Economy: Public Policy and Economic Change in America, 1900–1933.* Cambridge, MA: Harvard University Press, 1990.

Kettl, Donald. *The Transformation of Governance: Public Administration for Twenty-First Century America.* Baltimore: Johns Hopkins University Press, 2002.

Kick, John. "The Railway Telegraph Department." *Telephone Engineer* 5, (5), (May 1911).

Kieve, Jeffrey. *The Electric Telegraph: A Social and Economic History.* Newton Abbot: David & Charles, 1973.

Kingsbury, John. *The Telephone and Telephone Exchanges: Their Invention and Development.* New York: Arno Press, 1972 [1915].

Kingsbury, N. C. "Co-operation and the Telephone." *Telephone Engineer* 17, (3), (March 1917).

Kittler, Friedrich. "Thinking Colours and/or Machines." *Theory, Culture & Society* 23, (7/8), (2006): 39–50.

Klose, Nelson. *A Concise Study Guide to the American Frontier.* Lincoln: University of Nebraska Press, 1964.

Kozmetsky, George and Kircher, Paul. *Electronic Computers and Management Control.* New York: McGraw-Hill Book Co., 1956.

Krippendorf, Klaus. "The Systems Approach to Communication." In *General Systems Theory and Human Communication*, edited by Brent Ruben and John Kim. Rochelle Park: Hayden Book Co, 1975.

— "The Past of Communication's Hoped-For Future." *Journal of Communication* 43, (3), (Summer 1993). Reprinted in *Defining Media Studies*, edited by Mark Levy and Michael Gurevitch. New York: Oxford University Press, 1994.

Kurtz, John. "State Public Service Commissions." In Ernst Freund, Robert Fletcher, Joseph Davies, Cuthbert Pound, John Kurtz and Charles Nagel, *The Growth of American Administrative Law.* St. Louis: Thomas Law Book Co., 1923.

Kurtz, Lester. *Evaluating Chicago Sociology.* Chicago: University of Chicago Press, 1984.

Lacan, Jacques. "The Circuit." In *The Seminar of Jacques Lacan*, edited by Jacques-Alain Miller. New York: Norton, 1988.

Lardner, Dr. and Bright, Edward. *The Electric Telegraph*. London: James Walton, 1867.
Lasswell, Harold D. *Propaganda Technique in the World War*. London: Kegan Paul, Trench, Trubner & Co., 1938 [1927].
Lasswell, Harold. "The Structure and Function of Communication in Society." In *The Communication of Ideas*, edited by Lyman Bryson. New York: Institute for Religious and Social Studies, 1948. pp. 203–43.
Latil, Pierre de. *Thinking by Machine: A Study of Cybernetics*. Translated by Y. M. Golla. London: Scientific Book Guild, 1956.
Latour, Bruno. *The Pasteurization of France*. Cambridge, MA: Harvard University Press, 1988.
Lawson, R. "A History of Automatic Telephony." *Post Office Electrical Engineers' Journal* 5, (1912–13).
Leavitt, Harold and Whisler, Thomas. "Management in the 1980s." *Harvard Business Review* 36, (6), (1958): 41–8.
Lebow, Irwin. *Information Highways and Byways: From the Telegraph to the 21st Century*. New York: IEEE Press (1995).
Lee, John. "The Telephone and Social Reconstruction." *Telephone Engineer* 18, (2), (August 1917).
Lefferts, Marshall. "The Electric Telegraph: Its Influence and Geographical Distribution." *Bulletin of American Geographical and Statistical Society* 2, (January 1857).
Lekachman, Robert. *The Age of Keynes*. New York: Random House, 1966.
Leonard, V. A. *Police Communication Systems*. Berkeley: University of California Press, 1938.
— *The Police Communications System*. Springfield: Charles C. Thomas, 1970.
Leuchtenburg, William. *Franklin D. Roosevelt and the New Deal, 1932–1940*. New York: Harper & Row, 1963.
Levin, Howard. *Office Work and Automation*. New York: John Wiley & Sons, 1956.
Lilienfeld, Robert. *The Rise of Systems Theory: An Ideological Analysis*. New York: John Wiley & Sons, 1978.
Lindley, Lester. *The Constitution Faces Technology: The Relationship of the National Government to the Telegraph, 1866–1884*. New York: Arno Press, 1975.
— *The Impact of the Telegraph on Contract Law*. New York: Garland, 1990.
Lippman, Walter. *Drift and Mastery*. Englewood Cliffs: Prentice-Hall, 1961 [1914].
Lippmann, Walter. *The Good Society*. London: George Allen & Unwin, 1938.
Livermore, O. N. "Factory Telephones." *Telephone Engineer* 1, (2), (February 1909).
Locke, John. *An Essay Concerning Human Understanding*. Edinburgh: Mundell & Son, 1801 [1690].
Luhmann, Niklas. "How Can the Mind Participate in Communication?" In *Materialities of Communication*, edited by Hans Gumbrecht and K. Ludwig Pfeiffer. Translated by William Whobrey. Stanford: Stanford University Press, 1994.
Lustig, Jeffrey. *Corporate Liberalism: The Origins of Modern American Political Theory, 1890–1920*. Berkeley: University of California Press, 1982.
Lyytinen, Kalle. "A Taxonomic Perspective of Information Systems Development: Theoretical Constructions and Recommendations." In *Critical Issues in Information Systems Research*, edited by R. J. Boland Jr. and R. A. Hirschheim. Chichester: John Wiley & Sons, 1987.
Macbride, Robert. *The Automated State: Computer Systems as a New Force in Society*. Philadelphia: Chilton Book Co., 1967.
McCormick, John. *Carl Schmitt's Critique of Liberalism: Against Politics as Technology*. Cambridge: Cambridge University Press, 1997.
McLuhan, Marshall. *Understanding Media: The Extensions of Man*. New York: McGraw Hill, 1964.
McLuhan, Marshall and Fiore, Quentin. *The Medium is the Massage: An Inventory of Effects*. New York: Bantam, 1967.
McLuhan, Marshall and Powers, Bruce. *The Global Village: Transformations in the World Life and Media in the 21st Century*. New York: Oxford University Press, 1989.
Macrae, Donald. "Cybernetics and Social Science." *British Journal of Sociology* 2, (2), (1951): 135–49.
"Making the Public Efficient." *Telephone Engineer* 23, (4), (April 1920).
"The March of the Telephone." *Manchester Guardian*, October 5, 1923.
Marcus, Alan and Segal, Howard. *Technology in America: A Brief History*. San Diego: Harcourt Brace Jovanovich, 1989.
Maron, M. E. "On Cybernetics, Information Processing, and Thinking." In *Cybernetics of the Nervous System*, edited by Norbert Wiener and J. P. Schadé. *Progress in Brain Research*, vol. 17. Amsterdam: Elsavier, 1965.
Marshall, Margaret. "'Automation' Today and in 1662." *American Speech* 32, (2), (1957): 149–51.
Martin, Michèle. *"Hello, Central?" Gender, Technology, and Culture in the Formation of Telephone Systems*. Montreal: McGill-Queen's University Press, 1991.

Martin, Reinhold. "The Organizational Complex: Cybernetics, Space, Discourse." *Assemblage* 37, (1998): 102–27.
Marvin, Carolyn. *When Old Technologies Were New: Thinking about Electric Communication in the Late Nineteenth Century.* New York: Oxford University Press, 1988.
Mason, Richard and Mitroff, Ian. "A Program for Research on Management Information Systems." *Management Science* 19, (5), (1973): 475–87.
Matthews, Fred. *Quest for an American Sociology: Robert E. Park and the Chicago School.* Montreal: McGill-Queen's University Press, 1977.
Matulenas, Raymond. *Communication: A Source of Privileges.* Dissertation, Catholic University of America. Canon Law Studies No. 183. Washington, DC: Catholic University of America Press, 1943.
Maturana, Humberto and Varela, Francisco. *Autopoiesis and Cognition: The Realization of the Living.* Dordrecht: D. Reidel, 1980.
Mayr, Otto. *Authority, Liberty and Automatic Machinery in Early Modern Europe.* Baltimore: Johns Hopkins University Press, 1986.
Mead, George Herbert. "Review of C. L. Morgan: An Introduction to Comparative Psychology." *Psychological Review* 2, (July 1895).
Meier, R. L. "Automatism in the American Economy." *Journal of Business* 29, (1), (1956): 14–27.
Meinig, D. W. *The Shaping of America: A Geographical Perspective on 500 Years of History; Vol. 2, Continental America, 1800–1867.* New Haven: Yale University Press, 1993.
Melossi, Dario. *The State of Social Control: A Sociological Study of Concepts of State and Social Control in the Making of Democracy.* Worcester: Polity Press, 1990.
Meyrowitz, Joshua. *No Sense of Place: The Impact of Electronic Media on Social Behavior.* New York: Oxford University Press, 1985.
Miller, Char. *Taylored Citizenship: State Institutions and Subjectivity.* Westport, CT: Praeger, 2002.
Miller, Peter and O'Leary, Ted. "Hierarchies and American Ideals, 1900–1940." *Academy of Management Review* 14, (2), (1989): 250–65.
Mills, John. *Signals and Speech in Electrical Communication.* New York: Harcourt, Brace & Co., 1934.
Mindell, David. *Between Human and Machine: Feedback, Control, and Computing before Cybernetics.* Baltimore: Johns Hopkins University Press, 2002.
Mingers, John. *Self-Producing Systems: Implications and Applications of Autopoiesis.* New York: Plenum Press, 1995.
Mintzberg, Henry. *The Rise and Fall of Strategic Planning.* London: FT Prentice Hall, 2000.
Mirowski, Philip. *Machine Dreams: Economics Becomes a Cyborg Science.* Cambridge: Cambridge University Press, 2002.
Morgan, Gareth. *Images of Organization.* Beverly Hills: Sage, 1986.
Morse, Philip and Kimball, George. *Methods of Operations Research.* New York: John Wiley & Sons, 1951.
Morse, Samuel F. B. *His Letters and Journals: Invention of the Telegraph, Vol. II,* edited and supplemented by Edward Lind Morse. Boston: Houghton Mifflin, 1914.
Mott, Frank. *The News in America.* Cambridge, MA: Harvard University Press, 1952.
Mueller, Milton, Jr. *Universal Service: Competition, Interconnection, and Monopoly in the Making of the American Telephone System.* Cambridge: MIT Press, and Washington, DC: AEI Press, 1997.
Mulgan, G. J. *Communication and Control: Networks and the New Economies of Communication.* Cambridge: Polity Press, 1991.
Mumford, John. "This Land of Opportunity: The Nerve-Centre of Modern Business." *Harper's Weekly* (August 1, 1908).
Murdock, Graham. "Communications and the Constitution of Modernity." *Media, Culture and Society* 15, (1993): 521–39.
Murphy, Chandler. "Regulation of Service." *Telephone Engineer* 23, (4), (April 1920).
Nagel, Paul. *The Sacred Trust: American Nationality 1798–1898.* New York: Oxford University Press, 1971.
Nagle, James. *A History of Government Contracting.* Washington, DC: George Washington University, 1992.
Nancy, Jean-Luc. *The Birth of Presence.* Stanford: Stanford University Press, 1993.
— *Being Singular Plural.* Translated by Robert Richardson and Anne O'Byrne. Stanford: Stanford University Press, 2000.
Neering, Rosemary. *Continental Dash: The Russian-American Telegraph.* Ganges: Horsdal & Schubart, 1989.
Neisser, Ulric. "Computers as Tools and as Metaphors." In *The Social Impact of Cybernetics,* edited by Charles Dechert. Notre Dame: University of Notre Dame Press, 1966.
Neuman, Russell, McKnight, Lee, and Solomon, Richard. *The Gordian Knot: Political Gridlock on the Information Highway.* Cambridge: MIT Press, 1997.

"A New Police Signal." *Scientific American* (September 28, 1895).
"Newspaper Service by Telephone." *Telephone Engineer* 7, (3), (March 1912).
Noble, David. *America by Design: Science, Technology, and the Rise of Corporate Capitalism*. New York: Alfred A. Knopf, 1977.
Nye, David. "Shaping Communication Networks: Telegraph, Telephone, Computer." *Social Research* 64, (3), (1997).
Nye, Russel. *This Almost Chosen People: Essays in the History of American Ideas*. East Lansing: Michigan State University Press, 1966.
Nyquist, Harry. "Certain Factors Affecting Telegraph Speed." *Bell System Technical Journal* 3, (April 1924): 324–46.
Orlikowski, Wanda and Baroudi, Jack. "Studying Information Technology in Organizations: Research Approaches and Assumptions." *Information Systems Research* 2, (1), (1991): 1–28.
Otis, Laura. *Networking: Communicating with Bodies and Machines in the Nineteenth Century*. Ann Arbor, MI: University of Michigan Press, 2001.
Page, Arthur. "Social Aspects of Communication Development." *Bell Telephone Quarterly* 11, (2), (1932).
— "Social Aspects of Communication Development." In *Modern Communication*, edited by Arthur Page, H. D. Arnold, John E. Otterson, Harvey Fletcher, Ralph Brown, Frank B. Jewett, and Herbert E. Ives. Boston: Houghton Mifflin (1932).
Paglin, Max, ed. *A Legislative History of the Communications Act of 1934*. New York: Oxford University Press, 1989.
Palmer, W. E. "The Function of the Signal Corps." *Telephone Engineer* 18, (1), (July 1917).
Park, Robert. *Society*. Glencoe, IL: Free Press, 1955.
— *On Social Control and Collective Behavior*, edited by Ralph Turner. Chicago: Phoenix Books, University of Chicago Press, 1967.
— *The Crowd and the Public and Other Essays*, edited by Henry Elsner Jr. Chicago: University of Chicago Press, 1972.
Park, Robert and Burgess, Ernest. *Introduction to the Science of Sociology*. Chicago: University of Chicago Press, 1924 [1921].
Parsons, Talcott. *The Structure of Social Action*. New York: Free Press, 1949.
Parsons, Theophilus. *Law of Contracts*. Boston: Little, Brown, 1857 [1855].
Pearson, Keith. *Germinal Life: The Difference and Repetition of Deleuze*. New York: Routledge, 1999.
Peritz, Rudolph. *Competition Policy in America, 1888–1992: History, Rhetoric, Law*. New York: Oxford University Press, 1996.
Peters, John. "Satan and Savior: Mass Communication in Progressive Thought." *Critical Studies in Mass Communication* 6, (1989): 247–63.
Philips, William and Robinson, Charles. *Philips' & Robinson's Municipal Telegraph*. No publisher information, 1857.
Pickering, Andrew. "Cybernetics and the Mangle: Ashby, Beer and Pask." *Social Studies of Science* 32, (3), (2002): 413–37.
— "The Science of the Unknowable: Stafford Beer's Cybernetic Informatics." *Kybernetes* 33, (3/4), (2004): 499–521.
Pierce, J. R. *Symbols, Signals and Noise: The Nature and Process of Communication*. London: Hutchinson of London, 1962.
Pierce, John. "The Telephone and Society in the Past 100 Years." In *The Social Impact of the Telephone*, edited by Ithiel de Sola Pool. Cambridge: MIT Press, 1977.
Plum, William. *The Military Telegraph During the Civil War in the United States, Vol. I*. Chicago: Jansen, McClurg & Co., 1882.
Pool, Ithiel de Sola, ed. *The Social Impact of the Telephone*. Cambridge: MIT Press, 1977.
Pool, Ithiel de Sola. *Forecasting the Telephone: A Retrospective Technology Assessment*. Norwood: Ablex, 1983.
Pound, Arthur. *The Telephone Idea: Fifty Years After*. New York: Greenberg, 1926.
Pound, Roscoe. "Liberty of Contract." *Yale Law Journal* 18 (1909).
Pred, Allan. *The Spatial Dynamics of U.S. Urban-Industrial Growth, 1800–1914: Interpretive and Theoretical Essays*. Cambridge: MIT Press, 1966.
— *Urban Growth and City-Systems in the United States, 1840–1860*. Cambridge, MA: Harvard University Press, 1980.
Prescott, George. *History, Theory, and Practice of the Electric Telegraph*. Boston: Ticknor & Fields, 1860.
— *Bell's Electric Speaking Telephone: Its Invention, Construction, Application, Modification and History*. New York: D. Appleton & Co., 1884. Reprinted New York: Arno Press, 1972.

Pupin, Michael. *The New Reformation: From Physical to Spiritual Realities.* New York: Charles Scribner's Sons, 1927.

Pursell, Carroll. *The Machine in America: A Social History of Technology.* Baltimore: Johns Hopkins University Press, 1995.

Quandt, Jean. *From the Small Town to the Great Community: The Social Thought of Progressive Intellectuals.* New Brunswick: Rutgers University Press, 1970.

"The Railroad and the Telephone." *Telephone Engineer* 3, (1), (January 1910).

Rav, Yehuda. "Perspectives on the History of the Cybernetics Movement: The Path to Current Research through the Contributions of Norbert Wiener, Warren McCulloch, and John von Neumann." *Cybernetics and Systems: An International Journal* 33, (8), (2002): 779–804.

Raymond, Allen. *What is Technocracy?* New York: Whittlesey House, McGraw-Hill Book Co., 1933.

Recent Economic Changes in the United States. Report of the Committee on Recent Economic Changes, vol. 1. New York: McGraw-Hill Book Co., 1929.

Reid, James. *The Telegraph in America: Its Founders, Promoters and Noted Men.* New York: Derby Brothers, 1879.

Rhodes, Frederick. *Beginnings of Telephony.* New York: Harper & Brothers, 1929.

Robinson, Glen. "Title I, The Federal Communications Act: An Essay on Origins and Regulatory Purpose." In *A Legislative History of The Communications Act of 1934*, edited by Max Paglin. New York: Oxford University Press, 1989.

Rockefeller, Steven. *John Dewey: Religious Faith and Democratic Humanism.* New York: Columbia University Press, 1991.

Rogers, Everett. *Communication Technology: The New Media in Society.* New York: Free Press, 1986.

Rose, Nikolas. *Powers of Freedom: Reframing Political Thought.* Cambridge: Cambridge University Press, 1999.

Rosenberg, Charles. *No Other Gods: On Science and American Social Thought.* Baltimore: Johns Hopkins University Press, 1976.

Rosenblueth, Arturo, Wiener, Norbert, and Bigelow, Julian. "Behavior, Purpose and Teleology." *Philosophy of Science* 10, (1), (1943): 18–24.

Rosenblueth, Arturo and Wiener, Norbert. "Purposeful and Non-Purposeful Behavior." *Philosophy of Science* 17, (4), (1950): 318–26.

Ross, Dorothy. "The Development of the Social Sciences." In *The Organization of Knowledge in Modern America, 1860–1920*, edited by Alexandra Oleson and John Voss. Baltimore: Johns Hopkins University Press, 1979.

— *The Origins of American Social Science.* Cambridge: Cambridge University Press, 1991.

— "Modernist Social Science in the Land of the New/Old." In *Modernist Impulses in the Human Sciences, 1870–1930*, edited by Dorothy Ross. Baltimore: Johns Hopkins University Press, 1994.

Ross, Edward. *Social Control.* New York: Macmillan, 1901.

Rossides, Daniel. "William Graham Sumner (1840–1910)." In *The Classical Tradition in Sociology: The American Tradition*, vol. I, edited by Jeffrey Alexander, Raymond Boudon and Mohamed Cherkaoui. London: Sage, 1997.

Rowsey, A. "Telephones in the New York Police Bureau." *Telephone Engineer* 2, (1), (July 1909).

Roy, William. *Socializing Capital: The Rise of the Large Industrial Corporation in America.* Princeton: Princeton University Press, 1997.

Runde, Jochen and Mizuhara, Sohei, eds. *The Philosophy of Keynes's Economics: Probability, Uncertainty and Convention.* London: Routledge, 2003.

Russel, Robert. *Improvement of Communication with the Pacific Coast as an Issue in American Politics, 1783–1864.* Cedar Rapids: Torch Press, 1948.

Sarnoff, David. "Electronic Revolution, Present and Future." *New York Times*, September 30, 1956.

Schäffle, Albert. *Bau und Leben des socialen Körpers.* Tübingen: Verlag der H. Lauppschen Buchhandlung, 1881.

Schiller, Dan. "Social Movement in Telecommunications: Rethinking the Public Service History of US Telecommunications, 1894–1919." *Telecommunications Policy* 22, (4/5), (1998): 397–408.

Schmitt, Carl. *The Leviathan in the State Theory of Thomas Hobbes: Meaning and Failure of a Political Symbol.* Translated by George Schwab and Erna Hilfstein. Westport, CT: Greenwood Press, 1996.

Schwartz, Leonard. "Communication and the Computing Machine." In *The Evolving Society: The Proceedings of the First Annual Conference on the Cybercultural Revolution; Cybernetics and Automation*, edited by Alice Hilton. New York: Institute for Cybercultural Research, 1966.

Schwarzlose, Richard. *The Nation's Newsbrokers: Vol. 1, The Formative Years, from Pretelegraph to 1865.* Evanston: Northwestern University Press, 1989.

— *The Nation's Newsbrokers: Vol. 2, The Rush to Institution, from 1865 to 1920*. Evanston: Northwestern University Press, 1990.
Schwendinger, Herman and Schwendinger, Julia. *The Sociologists of the Chair: A Radical Analysis of the Formative Years of North American Sociology, 1883–1922*. New York: Basic Books, 1974.
Scott, Bernard. "Second-Order Cybernetics: An Historical Introduction." *Kybernetes* 33, (9/10), (2004): 1365–78.
Scott, Howard. "Technology Smashes the Price System." *Harper's Magazine* 166, (January 1933).
— "America Prepares for a Turn in the Road." *Technocracy* 1, (7), (1935).
Scott, Howard and others. *Introduction to Technocracy*. London: John Lane, The Bodley Head, 1933.
Sears, Vinton, ed. *Telephone Development: Status of the Industry, Scope and Effect of Competition*. Boston: Barta Press, 1905.
Serres, Michel. *Le contrat naturel*. Éditions François Bourin, 1990 (*The Natural Contract*. Translated by Elizabeth MacArthur and William Paulson. Ann Arbor, MI: Michigan University Press, 1995).
Seward, William. *Communication of Hon. William H. Seward, Secretary of State, upon the Subject of an Intercontinental Telegraph Connecting the Eastern and Western Hemispheres by way of Behring's Strait*. Washington: Government Printing Office, 1864.
Shaffner, Tal. *The Telegraph Manual*. New York: Pudney & Russell, 1859.
Shannon, Claude. "A Mathematical Theory of Communication." *Bell System Technical Journal* 27, (3), (1948): 379–423.
Shannon, Claude and Weaver, Warren. *The Mathematical Theory of Communication*. Urbana: University of Illinois Press, 1949.
Shapiro, Andrew. *The Control Revolution*. New York: PublicAffairs, 1999.
Sharlin, Harold. *The Making of the Electrical Age: From the Telegraph to Automation*. London: Abelard-Schuman, 1963.
Shaw, Thomas. "The Conquest of Distance by Wire Telephony." *Bell System Technical Journal* 23, (4), (1944).
Silbey, Joel. "Conclusion." In *A Crisis of Republicanism: American Politics in the Civil War Era*, edited by Lloyd Ambrosius. Lincoln: University of Nebraska Press, 1990.
Simmel, Georg. *The Sociology of Georg Simmel*, edited by Kurt Wolff. Glencoe, IL: Free Press, 1950.
Simon, Herbert. *Administrative Behavior: A Study of Decision-Making Processes in Administrative Organizations* (4th edn). New York: Free Press, 1997 [1945].
Simon, Herbert. *The Shape of Automation for Men and Management*. New York: Harper & Row, 1965.
Simon, Linda. *Dark Light: Electricity and Anxiety from the Telegraph to the X-Ray*. Orlando: Harcourt, 2004.
Sklansky, Jeffrey. *The Soul's Economy: Market Society and Selfhood in American Thought, 1820–1920*. Chapel Hill: University of North Carolina Press, 2002.
Sklar, Martin. *The Corporate Reconstruction of American Capitalism, 1890–1916: The Market, the Law, and Politics*. Cambridge: Cambridge University Press, 1988.
Skowronek, Stephen. *Building A New American State: The Expansion of National Administrative Capacities, 1877–1920*. Cambridge: Cambridge University Press, 1982.
Slesinger, Reuben. "The Pace of Automation: An American View." *Journal of Industrial Economics* 6, (3), (1958): 241–61.
Smith, Culver. *The Press, Politics, and Patronage: The American Government's Use of Newspapers, 1789–1875*. Athens: University of Georgia Press, 1977.
Smythe, Dallas. *The Structure and Policy of Electronic Communication*. Urbana: University of Illinois, 1957.
Snell, Edgar. "Factory Cost Accounting by Telephone." *Telephone Engineer* 1, (5), (May 1909).
Soule, George. *A Planned Society*. New York: Macmillan, 1932.
Spencer, Herbert. *Principles of Sociology* (3rd edn). New York: Appleton-Century-Crofts, 1897.
Standage, Tom. *The Victorian Internet*. London: Weidenfeld & Nicolson, 1998.
Startt, James and Sloan, William, eds. *The Significance of the Media in American History*. Northport: Vision Press, 1994.
Steinmetz, Charles. *America and the New Epoch*. New York: Harper, 1916.
Stern, Ellen and Gwathmey, Emily. *Once Upon a Telephone: An Illustrated Social History*. New York: Harcourt Brace & Co., 1994.
Stettner, Edward. *Shaping Modern Liberalism: Herbert Croly and Progressive Thought*. Lawrence: University Press of Kansas, 1993.
Stoikov, Vladimir. "Increasing Structural Unemployment Re-Examined." *Industrial and Labor Relations Review* 19, (3), (1966): 368–76.
Stokes, Melvyn and Conway, Stephen, eds. *The Market Revolution in America: Social, Political, and Religious Expressions, 1800–1880*. Charlottesville: University Press of Virginia, 1996.

Stone, Alan. *Public Service Liberalism: Telecommunications and Transitions in Public Policy*. Princeton: Princeton University Press, 1991.
Stroud, H. D. "Future Telephone Possibilities." *Telephone Engineer* 11, (1), (1914).
Sullivan, William. *Reconstructing Public Philosophy*. Berkeley: University of California Press, 1982.
Sumner, William. "Sociology." *Princeton Review* 57, (November 1881): 303–23.
— *What Social Classes Owe to Each Other*. Caldwell: Caxton Printers, 1978 [1883].
— *War and Other Essays*, edited by Albert Galloway Keller. New Haven: Yale University Press, 1911.
Susman, Warren. *Culture as History: The Transformation of American Society in the Twentieth Century*. New York: Pantheon Books, 1984.
Taksa, Lucy. "Scientific Management: Technique or Cultural Ideology?" Reprinted in *F. W. Taylor: Critical Evaluations in Business and Management*, vol. I, edited by John Wood and Michael Wood. London: Routledge, 2002.
Tarbell, Ida. *The Nationalizing of Business, 1878–1898*. New York: Macmillan, 1936.
Tariello, Frank, Jr. *The Reconstruction of American Political Ideology, 1865–1917*. Charlottesville: University Press of Virginia, 1982.
Tarr, Joel, Finholt, Thomas, and Goodman, David. "The City and The Telegraph: Urban Telecommunications in the Pre-Telephone Era." *Journal of Urban History* 14, (1), (1987): 38–80.
Taylor, Frederick. *The Principles of Scientific Management*. New York: W. W. Norton & Co., 1967 [1911].
— *Scientific Management: Comprising Shop Management, The Principles of Scientific Management, Testimony before the Special House Committee*, edited by Kenneth Thompson, vol. I of *The Early Sociology of Management and Organizations*. London: Routledge, 2003.
Taylor, George. *The Transportation Revolution, 1815–1860*. New York: Rinehart & Co., 1951.
Technocracy Study Course. Technocracy Inc., 1934.
Teeven, Kevin. *A History of the Anglo-American Common Law of Contract*. New York: Greenwood Press, 1990.
"Telephone Chief Wants Gossip Prohibited." *Telephone Engineer* 23, (4), (April 1920).
"The Telephone in Government Service." *Telephony* (June 1905).
"The Telephone in Modern Business." *Telephony* (November 1901).
Theobald, Robert. "Cybernetics and the Problems of Social Reorganization." In *The Social Impact of Cybernetics*, edited by Charles Dechert. Notre Dame: University of Notre Dame Press, 1966.
Thompson, John. *The Media and Modernity: A Social Theory of the Media*. Cambridge: Polity Press, 1995.
Thompson, Robert. *Wiring a Continent: The History of the Telegraph Industry in the United States, 1832–1866*. Princeton: Princeton University Press, 1947.
Tobey, Ronald. *The American Ideology of National Science, 1919–1930*. Pittsburgh: University of Pittsburgh Press, 1971.
Towers, Walter. *Masters of Space*. New York: Harper & Brothers, 1917.
Trafton, E. E. "Train Dispatching by Automatic Telephones." *Telephone Engineer* 15, (6), (June 1916).
"Train Orders by Telephone." *Telephone Engineer* 3, (3), (March 1910).
Tugwell, Rexford. *Industry's Coming of Age*. New York: Harcourt, Brace & Co., 1927.
Turnbull, Laurence. *Lectures on the Electro-Magnetic Telegraph*. Philadelphia: R. W. Barnard, 1852.
"Using the Telephone." *Telephone Engineer* 24, (4), (April 1921).
Vail, Alfred. *The American Electro Magnetic Telegraph with the Reports of Congress, and a Description of all Telegraphs Known, Employing Electricity or Galvanism*. Philadelphia: Lea & Blanchard, 1845.
Vail, Theodor. "Public Utilities and Public Policy." *Atlantic Monthly* 3, (1913).
Valentine, Herbert. "Social Efficiency." *Transactions of the Efficient Society* 1, (1912).
Valovic, Thomas. *Corporate Networks: The Strategic Use of Telecommunications*. Boston: Artech House, 1993.
Vattimo, Gianni. *The Transparent Society*. Translated by David Webb. Cambridge: Polity Press, 1992.
Veblen, Thorstein. *The Engineers and the Price System*. New York: Viking, 1924 [1919].
Vig, Norman. "Technology, Philosophy, and the State: An Overview." In *Technology and Politics*, edited by Michael Kraft and Norman Vig. Durham: Duke University Press, 1988.
von Neumann, John. *The Computer and the Brain*. New Haven: Yale University Press, 2000 [1958].
Waldo, Dwight. *The Administrative State: A Study of the Political Theory of American Public Administration* (2nd edn). New York: Holmes & Meier, 1984.
Ward, Lester. *Dynamic Sociology*, 2 vols. New York: Johnson Reprint, 1968 [1883].
— "The Claims of Political Science." In Lester Ward, *Glimpses of the Cosmos*, 6 vols. New York: Putnam, 1913–18 [1884].
Waring, Stephen. *Taylorism Transformed: Scientific Management Theory since 1945*. Chapel Hill: University of North Carolina Press, 1991.
Wasserman, Neil. *From Invention to Innovation: Long-Distance Telephone Transmission at the Turn of the Century*. Baltimore: Johns Hopkins University Press, 1985.

Webb, Herbert. "The Development of the Telephone in Europe." In *The Telephone Systems of the Continent of Europe*, edited by A. R. Bennett. New York: Arno Press, 1974.
Weber, Max. *From Max Weber: Essays in Sociology*, edited by Hans Gerth and C. Wright Mills. London: Routledge, 1948.
Weigley, Russell. *History of the United States Army*. New York: Macmillan, 1969.
Weizenbaum, Joseph. *Computer Power and Human Reason: From Judgement to Calculation*. San Francisco: W. H. Freeman, 1976.
— "Computers, Tools, and Human Reason." In *Communication in History: Technology, Culture, Society*, edited by David Crowley and Paul Heyer. New York: Longman, 1991.
Westbrook, Robert. *John Dewey and American Democracy*. Ithaca: Cornell University Press, 1991.
Whisler, Thomas and Shultz, George. "Automation and the Management Process." *Annals of the American Academy of Political and Social Science* 340, (1), (1962): 81–9.
White, Leonard. *The Jacksonians: A Study in Administrative History, 1829–1861*. New York: Macmillan, 1954.
— *The Republican Era: 1869–1901; A Study in Administrative History*. New York: Macmillan, 1958.
Wiebe, Robert. *The Search of Order, 1877–1920*. London: Macmillan, 1967.
— *Self-Rule: A Cultural History of American Democracy*. Chicago: University of Chicago Press, 1995.
Wiener, Norbert. *The Extrapolation, Interpolation, and Smoothing of Stationary Time Series*. Report 370, February 1, 1942. Records of NDRC, Sec 2, Div D. Washington, DC: National Archives & Records Service, 1942.
— "Time, Communication, and the Nervous System." *Annals* 50, (1948): 197–219.
— *The Human Use of Human Beings: Cybernetics and Society*. New York: Da Capo Press, 1954.
— *I am a Mathematician*. Cambridge: MIT Press, 1956.
— *Cybernetics, or, Control and Communication in the Animal and the Machine*. New York: MIT Press, 1961 [1948].
— *God and Golem, Inc.: A Comment on Certain Points where Cybernetics Impinges on Religion*. Cambridge: MIT Press, 1964.
Wiesner, Jerome. "Technology and the Telephone." In Arthur Clarke, Michael Dertouzos, M. Halle, I. de Sola Pool and J.B. Wiesner, *The Telephone's First Century—and Beyond*. New York: Thomas Y. Crowell, 1977.
Willey, Malcolm and Rice, Stuart. *Communication Agencies and Social Life*. New York: McGraw-Hill Book Co., 1933.
Wilson, David. *The National Planning Idea in U.S. Public Policy: Five Alternative Approaches*. Boulder, CO: Westview Press, 1980.
Wilson, George. *The Progress of the Telegraph*. Cambridge: Macmillan, 1859.
Wit, Dirk de. *The Shaping of Automation: A Historical Analysis of the Interaction between Technology and Organization, 1950–1985*. Hilversum: Verloren, 1994.
Wolff, Frank. "Why Standardize?" *Telephone Engineer* 22, (1), (July 1919).
Wolff, W. A. "Guiding Trains by Telephone." *Telephone Engineer* 16, (3), (September 1916).
Woodbury, David. *Communication*. New York: Dodd, Mead & Co., 1931.
Woods, David. *A History of Tactical Communication Techniques*. New York: Arno Press, 1974 [1965].
Wren, Daniel. *The History of Management Thought* (5th edn). Hoboken: John Wiley & Sons, 2005.
Yates, JoAnne. "The Telegraph's Effect on Nineteenth Century Markets and Firms." *Business and Economic History*, Series 2, 15, (1986).
— *Control through Communication: The Rise of System in American Management*. Baltimore: Johns Hopkins University Press, 1989.
— "Business Use of Information and Technology during the Industrial Age." In *A Nation Transformed by Information: How Information Has Shaped the United States from Colonial Times to the Present*, edited by Alfred Chandler Jr. and James Cortada. New York: Oxford University Press, 2000.
Young, John. *Cybernetics*. London: Iliffe Books, 1969.
Young, Peter. *Person to Person: The International Impact of the Telephone*. Cambridge: Granta Editions, 1991.
Zuboff, Shoshana. *In the Age of the Smart Machine: The Future of Work and Power*. Trowbridge: Heinemann, 1988.

Index